V

University of Hertfordshire

College Lane, Hatfield, Herts. AL10 9AB

Learning and Information Services
de Havilland Campus Learning Resources Centre, Hatfield

For renewal of Standard and One Week Loans,
please visit the web site **http://www.voyager.herts.ac.uk**

This item must be returned or the loan renewed by the due date.
The University reserves the right to recall items from loan at any time.
A fine will be charged for the late return of items.

Books by Gilbert Held, published by Wiley

Securing Wireless LANs
0 470 85127 9 (September 2003)

Ethernet Networks, 4th Edition
0 470 84476 0 (September 2002)

Quality of Service in a Cisco® Networking Environment
0 470 84425 6 (April 2002)

Bulletproofing TCP/IP-Based Windows NT/2000 Networks
0 471 49507 7 (April 2001)

Understanding Data Communications: From Fundamentals to Networking, 3rd Edition
0 471 62745 3 (October 2000)

High Speed Digital Transmission Networking: Covering T/E-Carrier Multiplexing, SONET and SDH, 2nd Edition
0 471 98358 6 (April 1999)

Data Communications Networking Devices: Operation, Utilization and LAN and WAN Internetworking, 4th Edition
0 471 97515 X (November 1998)

Dictionary of Communications Technology: Terms, Definitions and Abbreviations, 3rd Edition
0 471 97517 6 (May 1998)

Internetworking LANs and WANs: Concepts, Techniques and Methods, 2nd Edition
0 471 97514 1 (May 1998)

LAN Management with SNMP and RMON
0 471 14736 2 (September 1996)

virtual private networking

A Construction, Operation and Utilization Guide

GILBERT HELD

4-Degree Consulting, Macon, Georgia, USA

John Wiley & Sons, Ltd

This publication is designed to provide accurate and authoritative information in regard to the
subject matter covered. It is sold on the understanding that the Publisher is not engaged in
rendering professional services. If professional advice or other expert assistance is required, the
services of a competent professional should be sought.

Other Wiley Editorial Offices

John Wiley & Sons Inc., 111 River Street, Hoboken, NJ 07030, USA

Jossey-Bass, 989 Market Street, San Francisco, CA 94103-1741, USA

Wiley-VCH Verlag GmbH, Boschstr. 12, D-69469 Weinheim, Germany

John Wiley & Sons Australia Ltd, 33 Park Road, Milton, Queensland 4064, Australia

John Wiley & Sons (Asia) Pte Ltd, 2 Clementi Loop #02-01, Jin Xing Distripark, Singapore
129809

John Wiley & Sons Canada Ltd, 22 Worcester Road, Etobicoke, Ontario, Canada M9W 1L1

Wiley also publishes its books in a variety of electronic formats. Some content that appears
in print may not be available in electronic books.

British Library Cataloguing in Publication Data

A catalogue record for this book is available from the British Library

ISBN 0-470-85432-4

Typeset in 10.5/13pt Melior by Laserwords Private Limited, Chennai, India
Printed and bound in Great Britain by Biddles Ltd, Kings Lynn, Norfolk
This book is printed on acid-free paper responsibly manufactured from sustainable forestry
in which at least two trees are planted for each one used for paper production.

For longer than I care to admit I have been blessed with the opportunity to teach graduate school at Georgia College and State University. In doing so I am able to teach as well as learn from my students, a situation which has been extremely helpful for tailoring lectures and writing books and articles. In recognition of their indirect assistance, this book is dedicated to my students.

contents

preface

When we compare business operations today to years past, two key short phrases can be used to describe the difference between the two. Those phrases are 'mobile operations' and 'geographically separated offices.' Most businesses today employ telecommuters, have sales personnel who visit both customers and other corporate offices, and have technical workers whose destination on their next plane ticket may represent one of the great mysteries of life. When coupled with the need to interconnect geographically separated offices in an economical and secure manner, the requirements of business, government agencies and academia resulted in the development of hardware and software that can be used to create virtual private networks (VPNs) over public networks, which is the focus of this book.

In this book we will examine both the theory and practice of VPNs. Concerning the latter, we will examine, through the use of several series of screen displays, the creation of different types of VPNs. Because networking is not like clothing and the expression 'one size fits all' is not appropriate for solving the VPN requirements of different organizations, we will examine different types of VPNs from both a hardware and a software perspective. Through a detailed examination of the underlying technology associated with the construction and operation of VPNs it will be possible to compare and contrast the use of different protocols and authentication schemes, enabling us to make rational decisions concerning the use of different VPN components. Because economics represents one of the driving forces behind the use of VPN technology, we will also examine the cost associated with establishing several types of VPNs and then compare that cost to other networking alternatives. One of those alternatives will be the use of a VPN service, which could be suitable for organizations that do not have the personnel resources or time required to configure and manage the necessary hardware and software. Thus, it is the intent of this author to provide readers with both technical and financial information as well as networking options to facilitate your decision-making process.

As a professional author I highly value reader comments. Please feel free to contact me concerning any information presented in this book or information you think should be included in a future edition. Let me know if you feel I devoted too much space to a particular topic, if another topic needed additional elaboration or any other comments you may wish to share with me. You can send mail to me through my publisher whose address is on the jacket of this book or you can directly email me at *gil_held@yahoo.com*.

Gilbert Held
Macon, GA

acknowledgements

The publication of a book is a team effort that requires the knowledge and experience of many people beyond the name of the author displayed on the jacket. Thus, this author would be remiss if he did not acknowledge the contribution of many individuals who have made the publication of this book a reality.

The efforts of one's editors and publisher are instrumental in arranging reviews of an author's proposal, working with the author to orient and in some situations refocus the proposal, and selling the concept to the publisher's publication board. This author was fortunate to have Sally Mortimore and Birgit Gruber as Commissioning Editors to work with at John Wiley & Sons and Ann-Marie Halligan as Editorial Director.

Once a contract is approved, the interesting work begins. As an author that enjoys communicating hands-on information, doing so requires the setup of a computer laboratory. While my wife and Gizmo, my fine furry four-legged friend, have learned to put up with my various in-home cabling, I have finally migrated several of my computers to a wireless LAN environment. However, to be upfront with readers, I still use a large number of cables in my home office. Thus, while I can create a VPN from the laptop located in my kitchen to the server located in my office, when interconnecting two geographically separated hubs and servers in my office to simulate distant locations I had to perform a considerable amount of cabling. Both the hours spent cabling and the longer hours spent during the days and weekends I worked on this book reduced the amount of time I could spend with my wife. Once again, I appreciate her understanding as I labored and uttered 'expletives deleted' while configuring equipment.

As a frequent lecturer who periodically roams the globe, I have encountered a variety of electrical outlet receptacles, many of which refuse to mate with the adapters I would bring with me. After my laptop or notebook would beep due to an impending loss of battery power and with recharging not an option, I would turn to the use of my trusty pen and paper to write. After a few such travel episodes it was apparent that the pen and paper combination is mightier than the laptop or notebook and I began to draft my manuscripts the old-fashioned way. Using pen and paper requires a skilled typist or typists that can take this author's longhand draft and pencil illustrations and turn them into a professional manuscript. Once again I am indebted to Mrs. Linda Hayes and Mrs. Susan Corbitt for their fine work.

After a manuscript moves into production a considerable number of people are involved in editing, typesetting, cover design and other work. I would be remiss if I did not thank Daniel Gill for his fine effort in guiding my manuscript through the production process. Last but not least, I wish to thank everyone else involved in producing this book, to include many who literally work behind the scene but whose skills were equally important in producing the book you are now reading.

chapter one

Introduction to Virtual Private Networking

As with most books, we need an introduction to the topic, which is the purpose of this chapter. In this chapter we will become familiar with the concept behind virtual private networking, different types of VPNs you can find, the benefits and risks associated with the use of the technology, and alternatives that you may wish to consider. Concerning the latter, while the focus of this book is upon VPNs in a TCP/IP environment, this author would be remiss if he did not mention viable alternatives. In concluding this chapter we will take a brief tour of the contents of following chapters. You can use this tour by itself or in conjunction with the index and contents to locate one or more items of immediate interest. That said, grab a Coke, Pepsi or your favorite beverage and some munchies and follow me into the wonderful world of virtual private networks or VPNs.

1.1 The VPN concept

In commencing our initial discussion of VPNs it is probably wise to start with a definition of the technology. In doing so, let's describe both what a VPN is and also the term virtual private networking so we fully understand the difference between the two.

1.1.1 Definition

We can define a virtual private network as 'a temporary physical route formed over a mesh structured public network.' Thus, virtual private networking can

Virtual Private Networking   G. H. Held
© 2004 John Wiley & Sons, Ltd   ISBN: 0-470-85432-4

be considered to represent the process of transmitting data over a VPN. These basic definitions do not consider the layer in the Open System Interconnection (OSI) Reference Model at which the VPN operates nor does it describe any security requirements. As we will shortly note, the most popular type of VPNs operate at the data link layer (layer 2) or the network layer (layer 3). In addition, a less popular type of VPN, which we will briefly mention as well as a viable alternative to some layer 2 and layer 3 VPNs, operates at the transport layer (layer 4) while the key distribution protocol of a more recently standardized type of VPN operates at the session, presentation and application layers (layers 5 through 7). Because a VPN represents a temporary connection established over a public network, you more than likely want to secure your connection. Thus, the basic definitions previously presented represent a starting point or foundation for discussing VPNs.

1.1.2 Types of VPNs

There are two basic types of VPNs, those that operate at layer 2 in the OSI Reference Model and those that operate at layer 3.

1.1.2.1 Layer 2 VPNs

The top portion of Figure 1.1 illustrates a layer 2 VPN. In this example the layer 2 public network is a frame relay network. The frame relay network shown in Figure 1.1(a) is assumed to be a public network; however, it could also represent an organization's private network. Concerning the latter, while it is possible for an organization to establish a VPN over a private frame relay network, from a practical standpoint, it is probably illogical to do so. This is because a private frame relay network is formed by acquiring frame relay access devices (FRADs) and frame relay capable routers and switches and interconnecting those devices via leased lines. Since the network is in effect a private network, it makes little sense to create another private network over your own private network.

Returning to Figure 1.1(a), a frame relay network operates at layer 2 in the OSI Reference Model. The network consists of FRADs and routers or switches interconnected via leased lines. The VPN is established by configuring permanent virtual circuits (PVCs) where a PVC can be viewed as a portion of the capacity of a leased line. Thus, the VPN shown formed over the public frame relay network interconnects two remote locations at City A and City B to an organization's headquarters or central site located at City C. Note that the VPN can in this situation be considered as a fixed overlay to the typical mesh structure associated with public and private frame relay networks. Although

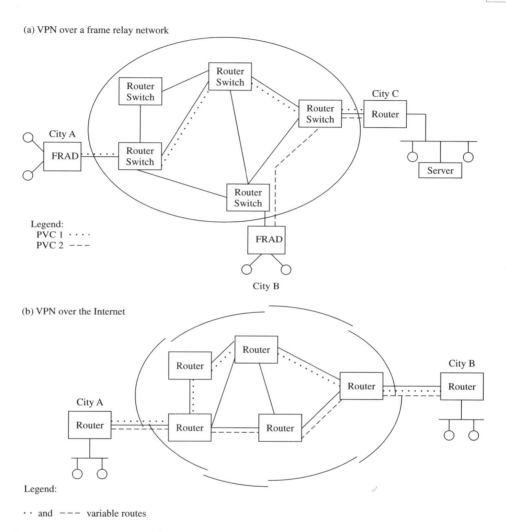

(a) VPN over a frame relay network

(b) VPN over the Internet

Figure 1.1 Many popular VPNs operate at either layer 2 or layer 3 of the OSI Reference Model.

the frame relay network can and normally is used to transport data from many organizations over its mesh-structured infrastructure, security is usually not an issue. This is because frame relay's permanent virtual circuits represent preconfigured routes through the network allocated for the use of predefined customers. Thus, unless the network operator does something he or she should not be doing, there is the minimal possibility for one customer of the network to observe the transmission of another customer. In addition, because PVCs are established on an end-to-end basis, there is normally no need to question the identity of the data originator. In comparison, when we turn our attention to the creation of a VPN over a layer 3 network, such as the Internet, we will note the need to verify who the originator claims to be via an established authentication method.

1.1.2.2 Layer 3 VPNs

As previously noted, layer 3 in the OSI Reference Model is the network layer. A VPN that operates at the network layer can occur over an X.25 network, a NetWare IPX network, a TCP/IP network, or a similar layer 3 network. In practicality, the only public network used for creating VPNs is the mother of all TCP/IP networks, the Internet.

Figure 1.1(b) illustrates the creation of a VPN over the Internet. In this example two geographically separated locations, again using the ubiquitous labels 'City A' and 'City B' to denote the locations, are connected via the mesh structure of the Internet. In comparing the VPN created over the frame relay network shown in Figure 1.1(a) to the VPN created over the Internet in Figure 1.1(b), the key difference between the two is in the manner by which data can flow. When the VPN is created over a frame relay network, the flow of data follows a fixed path since the VPN is created via the configuration of PVCs from source to destination. In comparison, when data flows over a layer 3 network, such as the Internet, routing can be considered to be variable since the path from source to destination is not only established on a temporary basis but can vary from packet to packet based upon the status of links between routers, router activity, the need of some traffic for expedited flowing on one path that precludes other traffic from using that path, and many other factors. Thus, Figure 1.1(b) illustrates the potential use of multiple paths between the geographically separated networks to support the flow of data between the two locations.

1.1.3 Categories of VPNs

There are two basic categories of VPNs – remote access and site-to-site. In this section we will examine the basic characteristics of each.

1.1.3.1 Remote Access VPN

A remote access VPN enables both fixed location and mobile workers to communicate with a central location. For example, a person working at home could have a DSL or cable modem connection to the Internet and require the ability to access one or more computers at the corporate headquarters or a regional office. If the corporate network includes a connection to the Internet, they could create a VPN connection from home to work. Similarly, an employee traveling to different locations may require the ability to access corporate computational facilities via the creation of a VPN from each location visited or even from a hotel room back to corporate headquarters. In doing so the traveling employee could dial the point of presence (POP) of a local Internet Service Provider (ISP) and establish a VPN connection to corporate headquarters, reducing the need for long-distance communications and a bank of telephone lines and modems at the home office to receive such calls.

Figure 1.2 illustrates two examples of remote access VPNs created over the Internet. At City A a home telecommuter is shown creating a VPN via a DSL modem connection to the Internet to corporate headquarters. At City B another employee of the same organization is using a dial-up connection to an ISP from a hotel room to create a VPN connection to the corporate network.

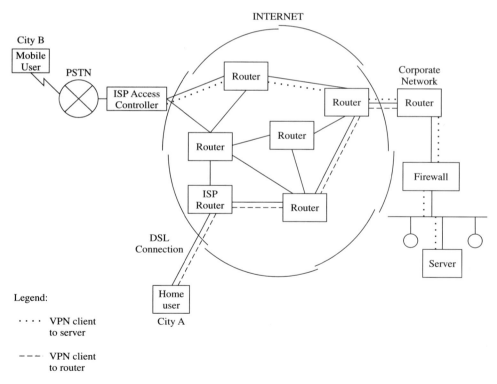

Figure 1.2   Remote access VPNs.

Remote access VPNs typically can terminate at a router or a network server, with the latter usually but not always providing additional network flexibility. In Figure 1.2 the home user's VPN is shown terminating at the corporate router that supports the VPN protocol used by the user. In comparison, the mobile user dialing an Internet Service Provider via the use of the public switched telephone network (PSTN) is shown creating a VPN to a server on the corporate network. Depending upon the VPN capability of the router, the client may be restricted to accessing a single computer or multiple computers connected to the corporate network. In a client-to-server VPN the configuration of the server would govern whether the client is restricted to accessing the server or can access other computers on the network via the server. Although the firewall can also be used to terminate a remote access VPN similar to the use of the router, it is shown for another reason. VPNs can be created using several protocols that will be described briefly in this introductory chapter and in much more detail later in this book. In the wonderful world of Internet communications many organizations employ firewalls as a security barrier to prevent unauthorized activities. While that is both reasonable and understandable, unless carefully examined and reconfigured, the firewall may block the creation of a VPN when its termination point is beyond the router.

Until the development of networking wizards that simplified the configuration process, the setup of a VPN client could be a most interesting experience. In addition, some of the encryption algorithms used by VPNs are processor intensive, which could degrade the performance of client computers when accessing a distant computer via the VPN or if they were attempting to perform other operations on the client during a VPN session.

In addition, the encryption algorithm performed on the router or server under the previously described remote access VPN scenarios would adversely effect equipment at the corporate network. This, as well as other issues that we will shortly discuss, resulted in the development of a second type of VPN topology referred to as a site-to-site VPN.

1.1.3.2 Site-to-Site VPN

A site-to-site VPN, as its name implies, interconnects two locations via a public network such as the Internet. A site-to-site VPN is commonly used to interconnect multiple clients at a branch office to corporate computers at a regional or headquarters location. In the reverse direction, clients at the regional or headquarters location have the ability to access computers at the branch office. Commonly, an organization might connect branch offices to regional offices and regional offices to headquarters through the establishment of multiple VPNs over the Internet. A company could also obtain a connection to another company across the Internet via the creation of a site-to-site VPN or a remote access VPN. In some literature this is referred to as an extranet VPN; however, in reality it is either a site-to-site or remote access VPN.

Figure 1.3 illustrates a site-to-site VPN. Note that the VPN functions are shown being performed by a gateway. The actual hardware used for a site-to-site VPN can include a router with a VPN capability, a firewall or a separate platform referred to as a VPN gateway. In examining Figure 1.3, note that each client at an organizational location simply accesses the VPN gateway and the gateway does the necessary work to create the VPN from one site to another. Thus, a site-to-site VPN relieves an administrator of the necessity to configure individual clients and also offloads authentication and encryption processing from those clients.

Returning to Figure 1.3, it is worth mentioning that many VPN gateways can be installed either before or behind the router that connects the site to

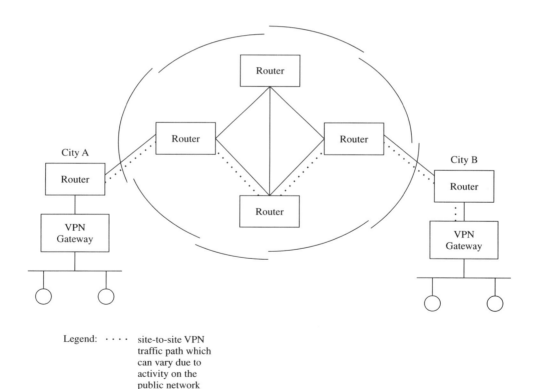

Legend: · · · · site-to-site VPN traffic path which can vary due to activity on the public network

Figure 1.3 Site-to-site VPN.

the Internet. The VPN gateway monitors network traffic and is transparent to packets flowing to other addresses than the address of the other site. When the gateway detects a destination address in a packet that is located at the other site, it performs its work, typically encrypting the packet as well as performing other functions that will be described later in this chapter as well as in more detail in subsequent chapters.

1.1.4 Infrastructure

In addition to categorizing VPNs by their type (remote access and site-to-site), we can also categorize this method of communications occurring over a public network by the operator of the network and the use of software or hardware to create the connection. Collectively, we will use the term infrastructure to reference the method of communications and the type of product used.

1.1.4.1 Method of Communications

Concerning the method of communications, VPNs can be established by an organization acquiring necessary hardware and/or software or through the use of a service contract. Concerning the latter, the contracted vendor is normally an ISP who either installs hardware at each user location for the creation of a site-to-site VPN or installs and directly controls hardware at the termination of the access line linking the customer to the ISP. Examples of communications vendors that provide VPNs as a service include AT&T, Worldcom, now renamed MCI after creating the largest accounting fraud in history, Infonet and Genuity, with the latter acquired by Level 3 Communications in early 2003.

If your organization already has the necessary infrastructure, such as VPN-capable hardware and trained personnel, you would probably elect to create the VPN. Otherwise, the use of a service could prove to be more economical. In addition, because the service provider normally has a large staff knowledgeable in the VPN configuration process, you more than likely do not have to worry about on-call support, vacation scheduling and similar factors associated with operating your own VPN.

1.1.4.2 Hardware vs. Software

A second category under infrastructure involves the manner by which you construct your VPN. If you do not require much bandwidth for your VPN connections, you should consider a software-only solution. You would then use software on clients to establish a VPN connection to a server, using either the VPN capability of the server's operating system or an add-on module operating on the server. If a number of clients access the server to the point where its performance degrades, you should consider adding a hardware encryption accelerator. If you are interconnecting two or more locations and need to support a large number of clients via individual VPN connections to one or more central locations, a hardware-based VPN would probably be better suited for your needs. Using a router with a VPN accelerator or a VPN gateway would remove the burden on both clients and server.

1.1.5 Benefits of use

While the previously presented information provides us with a basic understanding of the types and infrastructure associated with VPNs, until now we have not described the rationale for its use. The rationale for the use of VPNs resides in the areas of cost savings or economics and network reliability and availability.

1.1.5.1 Economics

To obtain an appreciation of how we can benefit economically from the use of a VPN, let's compare the use of a small router-based network created using three locations to the use of the Internet. Figure 1.4(a) illustrates the three-site router-based network. If we assume each location is one thousand miles distant from the other two locations and the cost of a 1000-mile T1 lines is $4.00 per mile per month, then the monthly cost of communications for the three site router based lease line network becomes:

$$2,000 \text{ miles} * \$4/\text{mile}/\text{month} = \$8,000/\text{month}$$

Note that the router at City A requires two WAN ports, while the routers located in Cities B and C each only require one WAN port. Also note that the configuration shown in Figure 1.4(a) has no backup capability. That is, if the link from City A to City B fails, users on networks located at City A and City C cannot access computers on the network located at City B. Similarly, computers located on the network at City B would not be able to access computers located on the networks at City A and City C. If the T1 line connecting the routers between City A and City C should fail, a similar networking problem would occur, with computers connected to the network located at City C incapable of connecting to computers connected to the networks at City A or City B, while computers connected to the networks located at City A and City B would be incapable of connecting to computers connected to the network located in City C.

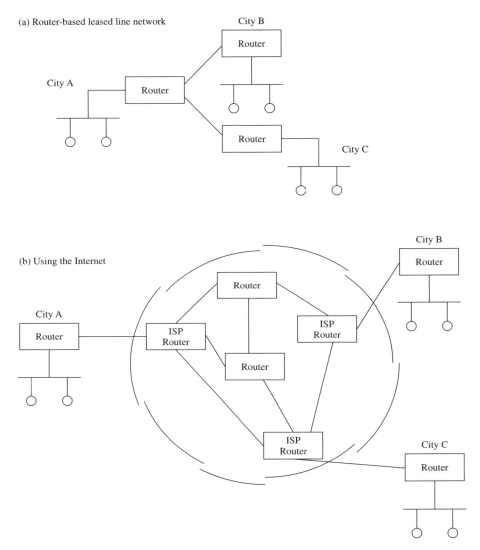

Figure 1.4   Comparing leased line and Internet-based networking.

While the installation of a T1 leased line between Cities B and C could be used to provide an alternate routing capability, doing so would increase both the monthly cost of communications and the cost of hardware. Concerning communications cost, again assuming City B is located 1000 miles from City C, the monthly cost of a T1 line connecting both locations would be

$4,000. Because the routers at Cities B and C would now each require two WAN ports, there would also be an additional hardware cost associated with obtaining the additional ports.

Turning our attention to the use of the Internet, let's discuss Figure 1.4(b). To use the Internet as a transmission system the router in each city is connected via an intracity T1 leased line to an ISP router. Thus, the cost of communications for each location is based upon the monthly cost of an intracity T1 circuit and the ISP's monthly fee for supporting a T1 connection to the Internet. Although costs will vary among ISPs, an organization can usually obtain a bundled package from an ISP to include an intracity T1 line and an Internet connection for approximately $2,000 per month. Thus, the cost associated with using the Internet to connect the three locations would become $6,000 per month.

Without considering the cost of hardware, you might be able to save $2,000 per month on a VPN solution as a mechanism to interconnect the three locations 1000 miles distant from one another. However, it is worthwhile to note that as the distance between locations increases, the potential savings via the use of a VPN correspondingly increase. This is because the leased line network cost is based upon the distance of the leased lines used to interconnect geographically separated locations while the intracity cost of an Internet connection is normally a fixed charge. For example, let's assume you wished to establish a network linking Seattle, Miami and New York, resulting in two circuits necessary to interconnect the three locations having approximately a length of 3,600 miles. At a cost of $4 per mile per month, the T1 circuits required to interconnect the three locations would have a monthly cost of $14,400. In comparison, the cost associated with using the Internet to interconnect the three locations would remain at $6,000 per month. Thus, as the distance between geographically separated locations increases, the potential savings associated with the use of a VPN to link those locations can also be expected to increase.

In examining the use of the Internet to connect the three locations, note that each router only requires one WAN port. Thus, there are hardware costs that can be reduced under this networking scenario that were not included in the previous computations but which should occur when performing a more detailed analysis. Now that we have an appreciation of the potential economic benefit associated with the use of VPNs, let's turn our attention to a second benefit associated with the use of this communications technology – redundancy.

1.1.5.2 Redundancy

The Internet infrastructure consists of a mesh network, with each router usually having alternate connections to other routers in the network. This means that there is a degree of built-in redundancy and alternate routing capability associated with using the Internet that requires additional cost to achieve when you construct a network through the use of leased lines. If you compare Figure 1.4(a) and 1.4(b) you note that when using the Internet there are typically multiple paths between routers. In addition, the mesh structure of the network permits the failure of one or more circuits or routers to be compensated for by the use of alternative paths. While this redundancy can be built into private networks, doing so can be cost-prohibitive for most organizations.

1.1.5.3 Network Flexibility and Scalability

While the prior comparison was of a site-to-site networking environment, a similar series of benefits are associated with the use of remote access VPN communications. Because they eliminate the need for long-distance telephone charges, remote access VPNs can reduce the cost of communications. Although it is usually good for remote access VPNs to reduce the communications costs, their primary benefit is one of networking flexibility. Once the client station obtains a VPN connection to a router, VPN gateway or server, they may be able to access different computers on the network via one connection instead of having to establish multiple connections. While both economics and flexibility are important, the use of the Internet also provides a degree of scalability usually not rapidly obtainable when constructing a private network unless you add transmission facilities beyond your immediate needs. While economics, reliability and scalability are of considerable importance, we would be remiss if we did not also discuss some of the disadvantages associated with the use of VPNs.

1.1.6 Disadvantages of VPNs

There are four major disadvantages associated with the use of VPNs. Of the four disadvantages listed, security is by far the major issue and is a governing factor for the other listed disadvantages. Thus, in this section we will primarily focus our attention upon VPN security and why it affects the other issues.

- Require an understanding of methods required to secure transmission over a public network.

- Can require custom configuration beyond the use of a wizard.
- Performance obtainable to include a quality of service (QoS) capability depends upon factors outside the control of the organization.
- Full interoperability between different vendor products may be difficult to achieve due to the complexity of some VPN-related standards.

1.1.6.1 Security

Because a VPN represents a connection established over a public network it is subject to a variety of security-related vulnerabilities. Those vulnerabilities can range from data being viewed by people using a network analyzer to other people who attempt to gain access to computational facilities by posing as legitimate clients. Due to such threats VPNs need encryption to hide the contents of their packets from observation as well as an authentication scheme to verify the identity of the remote user. In addition, an integrity checking mechanism is required to ensure that data is not tampered with during transmission from a 'man-in-the-middle' attack. Because the selection and distribution of encryption keys can represent a time-consuming activity as the number of VPN users increase or the number of site-to-site connections expands, many organizations require a key management system. Because knowledge is power, network managers and LAN administrators are well served by obtaining information about both legal and potentially illegal activities, such as invalid access attempts. This information can be provided by an accounting system. Last but certainly not least, a variety of configuration settings may be required to define varying connection permissions available to different users, a function referred to as authorization.

Table 1.1 lists what this author likes to refer to as the five pillars of network security that are applicable to VPNs and other types of communications. While each of the pillars is important, not all are supported by certain VPN protocols and, even when supported, an organization may elect to use a

TABLE 1.1 The pillars of network security

Function	Duty
Accounting	Record activities to note potential threats or illegal attempted activities
Authorization	Apply policy controls to network connectivity permissions
Encryption	Encode data to prevent unauthorized parties from viewing it
Integrity Check	Ensure data is not modified during transmission
Key Management	Select and distribute encryption keys

default setting of disabled if their requirements do not need to implement a particular security feature.

1.1.6.2 Configuration

Although the use of network wizards has certainly facilitated the VPN configuration process, it can still represent a time-consuming task. This is especially true when your organization has a sufficient number of network functions of which a subset is only permitted for many users.

1.1.6.3 Performance

When you use the Internet, many network-related functions and features may be placed beyond the control of your organization. For example, on a private network it is possible to configure the Resource Reservation Protocol (RSVP) and obtain a quality of service (QoS) that favors certain traffic over other types of traffic. When you create a VPN over the Internet, you may not be able to achieve a traffic expediting capability, especially if you use more than one ISP to provide your organization's Internet connections.

1.1.6.4 Interoperability

Although the vast majority of VPN technology is built upon networking standards, there are a significant number of features offered by different vendors to make interoperability an issue. While many VPN products from different vendors work well together, when you turn on certain features at one end of your VPN you may find that the client at the opposite end does not support that feature. Thus, you should carefully note your full requirements prior to considering equipment and software from multiple vendors.

Now that we have an appreciation of some of the disadvantages associated with the creation and utilization of VPNs, we will turn our attention to a brief overview of the protocols used by this communications technology.

1.1.7 VPN protocols

In this section we will briefly note the protocols that can be used to support the creation of VPNs. In doing so we will not discuss frame relay as the primary focus of this book is the creation of VPNs over the Internet. Thus, our discussion of VPN protocols will primarily be with respect to the Internet Protocol (IP) and layer 3 networking. This discussion is oriented to acquaint you with a brief overview of VPN protocols and specific information concerning many other protocols will be presented in considerably more detail later in this book.

1.1.7.1 PPP

The Point-to-Point Protocol (PPP) does not actually support VPNs, however, it is used as a communications foundation by some VPN protocols and therefore deserves mention. PPP represents a mechanism for transporting different network layer protocols, such as IPX and IP, over the same point-to-point communications link. PPP includes three major components. The first component is an encapsulating and decapsulating feature that provides the protocol with the ability to transmit and receive other network protocols. A second component is a Link Control Protocol (LCP) that is used for configuring, establishing, and testing the data-link connection. The third component is a family of Network Control Protocols (NCPs). The NCPs provides the mechanism for configuring and establishing different network layer protocols.

1.1.7.2 PPTP

The Point-to-Point Tunneling Protocol (PPTP) represents a mechanism that secures data transfer from a remote client to a network server via an IP network, such as the Internet. PPTP operates by creating a VPN through tunneling PPP frames.

PPTP uses two basic types of packets – data and control. Data packets are used to transmit user data and are encapsulated using the Generic Routing Encapsulation Protocol (GRE). Control packets flow in the clear and are used for signaling and status queries. A third type of packet, management, is defined in the PPTP specification but their contents have yet to be defined.

Both hardware and software vendors contributed to the development of PPTP. This resulted in two basic methods that can be employed to create a VPN through the Internet that will be described in detail later in this book. In addition, the method employed for authentication of users is left to the implementator but this can either provide the variability necessary to support emerging techniques or result in problems when attempting to integrate products from different vendors.

1.1.7.3 L2TP

The Layer 2 Tunneling Protocol (L2TP) evolved from an initial competitor to PPTP referred to as Layer 2 Forwarding (L2F). L2F was a protocol primarily implemented in Cisco Systems routers as a mechanism for creating User Datagram Protocol (UDP) encapsulated tunnels between remote access concentrators and routers. In an attempt to improve upon L2F as well as garner more support for a competitive method to PPTP, several vendors worked on combining the best features of L2F and PPTP, resulting in a new standard referred to as L2TP.

L2TP operates by tunneling data from an ISP access controller to a network server. Although L2TP is similar to PPTP in that it supports non-IP clients, it exceeds the capability of the former in that it extends VPN support to frame relay, Asynchronous Transport Mode (ATM) and even Synchronous Optical Network (SONET) networks. Because L2TP does not define an encryption method, one common method of securing L2TP is to use IPSec with L2TP. When L2TP is combined with PPTP, both data and control packets are encrypted. In comparison, under PPTP, only data packets are encrypted.

1.1.7.4 IPSec

Internet Protocol Security (IPSec) represents a collection of related protocols that provide authentication, encryption and key management capability. IPSec extends the standard Internet Protocol (IP) to provide a support for obtaining a secure Internet service that can include but is not limited to obtaining a VPN capability. In addition to client authentication, IPSec provides a data authentication mechanism that verifies the integrity of received data and prevents a 'man-in-the-middle' attack.

1.1.7.5 SOCKS

The acronym behind SOCKS represents a mystery to this author and it is doubtful if it had anything to do with President Clinton's cat as its development preceded the feline. SOCKS represents a protocol that operates at the session layer (layer 5) in the OSI Reference Model. As a layer 5 protocol SOCKS has the ability to enable network managers and LAN administrators to directly control the type of traffic that will flow over a VPN. While SOCKS can be implemented in software, it operates as a proxy service, controlling the flow of data by establishing a virtual circuit between client and host for one or more previously specified applications. Because of the operation of the proxy service, this action will lower overall client performance. In addition, servers may not be able to support a large number of SOCKS sessions. Because the use of SOCKS is minimal in comparison to other VPN methods, its technology will not be discussed further.

1.1.7.6 SSL and TLS

Secure Sockets Layer (SSL) and Transport Layer Security (TLS) represent two additional transport layer protocols that can be used to create secure communication between clients and servers. SSL and TLS are very similar to one another and widely used in browsers to protect the exchange of data between clients and hosts. Most Web browsers including Microsoft's Internet Explorer and Netscape's Navigator support both SSL and TLS.

Both SSL and TLS require the use of certificates. In fact, a server must authenticate itself to the client through a certificate-based technique during which certificates are exchanged, session keys are generated and an encryption method is mutually selected. The primary difference between SSL and TLS is in the area of key generation, with the TLS mechanism required to be used in a module that requires Federal Information Processing Standard (FIPS) validation.

Although in theory SSL and TLS could be used to protect any TPC/IP application, in reality, they are used primarily with the HyperText Transport Protocol (HTTP), the protocol almost exclusively used for Web pages. This prevents SSL and TLS supported by Web browsers from being able to directly access other applications, such as email and the file transfer protocol (FTP) in a secure manner. As we will note later in this book, through the use of certain vendor products it becomes possible to create a VPN tunnel via a Web browser that can securely access many applications at the distant location.

1.1.8 Summary

Figure 1.5 provides a comparison of the ISO Open System Interconnection (OSI) Reference Model, the TCP/IP protocol suite, and various VPN-related protocols that can be used to provide secure communications over a public network. In examining Figure 1.5 note that the Challenge Handshake Authentication Protocol (CHAP), Password Authentication Protocol (PAP) and Microsoft's version of CHAP (MS-CHAP) listed at the data link layer represent three authentication methods that will be described in more detail later in this book. Although IPSec is primarily a layer 3 protocol with two of its three major components operating at the network layer – Authentication Header (AH) and Encapsulating Security Payload (ESP) – a third protocol, referred to as the Internet Security Association and Key Management Protocol (ISAKMP) operates as an application and is shown as such.

ISO OSI Reference Model	TCP/IP Protocol Suite	VPN-Related Protocols
Application	Application	IPSec (ISAKMP)
Presentation		
Session		
Transport	Transport (TCP/UDP)	SOCKS SSL, TLS
Network Layer	Network Layer (Internet Protocol)	IPSec (AH, ESP)
Data Link Layer	Data Link Layer	L2TP L2FP CHAP, PAP, MS-CHAP

Figure 1.5 Comparing the relationship between VPN-related protocols, the ISO OSI Reference Model and the TCP/IP protocol suite.

1.1.9 Alternatives to VPNs

In concluding our discussion of VPNs, it is worthwhile reviewing alternatives, including leased lines to create a site-to-site VPN equivalency and the use of different communications facilities to create a remote access VPN equivalency. Remote access VPN equivalency can occur through the use of modem and Integrated Services Digital Network (ISDN) dial services. In addition, for businesses accessing other business locations, it may be possible to use one of several switched digital services, such as switched 56 Kbps, switched 64 Kbps, or switched 384 Kbps. Other possible alternatives to the creation of VPNs include the use of project collaboration software, messenger software, and some remote access programs that now include an encryption capability.

1.1.9.1 Leased Lines

The use of leased lines or analog or digital switched services provides with the ability to utilize the infrastructure of one or more communications carriers to interconnect locations on either a permanent (leased line) or temporary (switched analog or digital) basis. Unless your organization is a bank or insurance company or transmits classified information, there is usually no need to encrypt data when interconnecting locations via the use of leased lines. Similarly, because each location is fixed, there is usually no need to authenticate users. Thus, many security issues associated with the use of VPNs fall by the wayside when leased lines are used.

1.1.9.2 Switched Network Use

Because certain modem, ISDN and switched digital calls can be used to verify the originator by the origination telephone number (caller ID), a similar argument concerning encryption and authentication can be made. Thus, only when dialing through the facilities of a hotel PBX or a foreign (non-organizational) office would encryption and authentication be important considerations.

1.1.10 Economic issues

In addition to security issues associated with the use of leased lines and the switched network, you need to consider costs. As indicated earlier, leased lines are billed on a monthly basis based upon their operating rate and distance between connected locations. That is, the higher the data rate and longer the transmission distance, the higher the monthly fee. Thus, when geographically separated locations are a relatively short distance from one another, the monthly cost of a private leased line network becomes more economical in comparison to the use of the Internet. However, if your organization needs an Internet connection, then you must consider the cost associated with upgrading your intracity Internet connections to the cost of a private leased line network. In most situations the cost associated with upgrading your intracity Internet connection to accommodate VPN traffic will be less than the cost of a private leased line network.

If we turn our attention to the remote access VPN equivalence obtained through analog or digital dial-up services, the cost of such services depends upon transmission location, transmission distance, and for digital switched service, the data rate. Concerning transmission location, it is a well-known fact that dialing directly from a hotel room can result in a considerable surcharge to the actual cost of the bill. Other special cost considerations include the cost associated with using a telephone credit card, the distance to the called party that could result in a long-distance call, and the time of day when the call is made. Because different communications carriers have different fees and different discount plans, the actual cost associated with using dial-up as an alternative to a remote access VPN can be expected to vary between organizations for similar usage. However, in this introductory chapter we should note the simple fact that the cost of dial-up calls is based on duration. Thus, the more calls and the longer each call, the higher the cost of the call. As a general rule of thumb, as the number of remote users requiring access to distant computational facilities increases, the better the economic case for a remote access VPN.

1.1.11 Other alternatives

Three alternative options you can consider include project collaboration software, messenger programs and remote access software that enables one computer to take control of another. With the exception of remote access software, it may be difficult to protect the other two. That is, many project collaboration programs to include 'grayboards' and 'blackboard' type programs do not include an authentication or an encryption option. While suitable for an intranet, the use of project collaboration does not provide the flexibility to access predefined applications on multiple computers that VPNs can provide. Similarly, messenger programs commonly lack security and do not provide the ability to access application programs.

The third option, remote access software, can be considered a viable alternative to a VPN. This is because modern remote access software programs include support for security. However, because the remote access software program enables the remote user to take charge of the distant computer, its use can result in new security issues. If your organization only needs to support one or a handful of remote users with access to applications at a common location, the use of remote access software can represent a viable alternative to VPN hardware and software. However, as the number of remote access users increases, in general, traditional VPN solutions become more attractive. In concluding this brief examination of VPN alternatives, we can paraphrase General McArthur and note that 'there is no substitute for performing an analysis based upon the operating environment of your organization.'

1.2 Book preview

In concluding this introductory chapter we will take a brief tour of succeeding chapters in this book. You can use this information either by itself or in conjunction with the contents and index to locate items of an immediate interest. Although this book was written to be read in chapter order sequence, this author also realizes the need to locate information of immediate interest.

Thus, each chapter was written to be as independent as possible of preceding and succeeding chapters.

1.2.1 Understanding authentication and cryptology

In Chapter 2 we will focus our attention upon two key security pillars – authentication and encryption. As previously noted in this chapter, authentication provides us with a mechanism to verify the identity of a person. In actuality, authentication techniques may be employed to identify the location of the data originator, the device or hardware from which data originated as well as the person sending the data. In Chapter 2 we will primarily discuss the authentication of the data originator and techniques and protocols used to support this method of identification.

In the second part of Chapter 2 we will look at the fascinating field of cryptology. We will first note how digital encryption and decryption are performed. Using this information as a foundation, we will examine private and public key operations to include the authentication property of public key systems, the use of hash functions and digital certificates.

1.2.2 Understanding the TCP/IP protocol suite

While your first impression might be to say, 'Oh, no, not another TCP/IP tutorial,' by the time you have read Chapter 3 your impression will hopefully have changed. The aim of Chapter 3 is to briefly review the TCP/IP protocol suite with respect to the creation and operation of VPNs. Thus, after briefly examining the structure of the protocol suite we will turn our attention to the key metrics that can be used to define a particular VPN as well as enable its traffic to flow through routers, firewalls and other networking devices. We will examine the IP, TCP and UDP headers, focusing on certain fields in each header as well as reviewing IP addressing. In addition, we will also review the Internet Control Message Protocol (ICMP) as the settings of router access lists and firewalls can have a bearing upon the ability to perform certain types of diagnostic tests and support certain types of VPNs. In concluding Chapter 3 we will discuss the process of network address translation (NAT) and proxy servers as they can have a significant effect upon the operation of VPNs.

1.2.3 Layer 2 VPN techniques

In Chapter 4, layer 2 VPN techniques are discussed. Because an understanding of layer 2 VPN technology requires knowledge of the Point-to-Point Protocol (PPP), we will commence this chapter with a brief review of this protocol.

Then we will examine the Point-to-Point Tunneling Protocol (PPTP) and Layer 2 Forwarding (L2F), as well as the Layer 2 Tunneling Protocol (L2TP) and how security can be enhanced by combining IPSec with L2TP. The latter will serve as an introduction to Chapter 5, which deals with higher layer VPNs.

1.2.4 Higher layer VPNs

In Chapter 4 we described the rationale for combining IPSec with L2TP. In Chapter 5 we will examine IPSec in detail as well as other higher layer VPN creation methods, such as SSL and TLS.

1.2.5 VPN hardware and software

We can view the first five chapters in this book as primarily being focused upon the underlying technology used to create VPNs. In Chapter 6 we will change orientation and focus our attention upon a representative series of hardware and software products that can be used to create different types of VPNs. In doing so we will also note how certain types of hardware and software products can be used to provide different levels of functionality.

1.2.6 Service provider-based VPNs

Although many organizations prefer to construct, operate and manage their own virtual private network, other organizations may lack trained personnel or need to focus resources on other areas. Such organizations are candidates for service provider-based VPNs which is the focus of Chapter 7. In Chapter 7 we will examine the rationale for server-provided VPNs and the features associated with their use.

Due to the vast amount of information presented in this book, a VPN checklist is included as an appendix. This checklist can be used to verify that you have considered the numerous factors that need to be taken into account to obtain a secure and cost-efficient VPN that will satisfy the requirements of your organization.

chapter two

Understanding Authentication and Encryption

The main aim of this chapter is to become acquainted with two of the key pillars of security: authentication and encryption. In the first section we will discuss several common authentication methods. In doing so we will slightly put the 'cart before the horse' as we will describe authentication methods associated with certain protocols prior to discussing those protocols. However, because authentication methods can be treated as a separate entity, we do not have to be concerned about their operation prior to describing the protocols that support the actual transfer of information. In the second section of this chapter we will examine two main categories of encryption associated with VPNs: public and private key. In addition, we will also describe the role of digital certificates and their role in verifying the owner of public and private keys as well as their role in certain types of authentication.

2.1 Authentication

Authentication represents the process of identifying a user, user location or computer. In this section we will describe the operation and utilization of a variety of authentication methods. While our primary focus will be upon authentication methods commonly used with VPNs, it is important to note that our discussion is by no means comprehensive and there are many other authentication methods that are available apart from those discussed in this section.

Virtual Private Networking   G. H. Held
© 2004 John Wiley & Sons, Ltd   ISBN: 0-470-85432-4

2.1.1 Password authentication protocol

Perhaps the earliest method employed to authenticate a remote user is password-based authentication. In reality, as we will shortly note, there are several types of password-based authentication methods. Perhaps the first method to be developed was the one that associates a user-name or user-identity or identification (user-ID) to a password. One method for using this type of authentication scheme was standardized as the password authentication protocol.

The password authentication protocol (PAP) is defined in RFC 1334. PAP is based upon a two-way handshaking procedure. In this section we will examine its operation as well as discuss some of the limitations of this protocol.

2.1.1.1 Operation

Once a connection between a remote client and the called party is established, the originator of the call transmits a user-ID/password pair to the called party. The called party, which is normally a server, was previously configured with a password for each user-ID account and performs a lookup to verify the received data pair. If the data pair is verified, the called party returns an acknowledgement. Otherwise, the called party either terminates the connection or responds to the originator with an error message and provides another opportunity for the originator to authenticate itself. In the wonderful world of communications the originator is normally a remote client while the called party is normally a network server.

2.1.1.2 Limitations

One of the major limitations associated with PAP is the fact that passwords are transmitted in the clear. This means a third party could monitor the session initialization sequence and repeat the user-ID/password later to gain access to a server, a process referred to as playback.

Another limitation of PAP is the fact that many servers are configured to allow only a small number of erroneous sign-on attempts. Once that number has been reached, the server will lock out the remote client from future access attempts, either for a predefined period of time or until the server administrator clears the account. This lockout mechanism is designed to prevent so-called dictionary attacks in which an unauthorized third party attempts to gain access to a user account by trying every entry in a dictionary as the password for the account. Unfortunately, an unauthorized third party could simply record every user-ID and write a script to try a series of passwords for each account on a server. If the person does this at 8 a.m. on a Monday, workers could arrive and find out they are locked out of their server accounts, creating an administrative nightmare!

2.1.1.3 Windows Dial-Up Networking

You can easily view the use of PAP in many Microsoft Windows products. For example, double-click on the My Computer icon and then on the Dial-up Networking icon. Then, right-click to select Properties from the pop-up menu. You will then view a dialog box similar to the one shown in Figure 2.1.

Figure 2.1 illustrates the positioning of the Networking tab in the foreground of the Dial-Up Networking Properties dialog box. Note that the type of dial-up server is selected as Point-to-Point Protocol (PPP), which is common for Windows 2000/NT and Windows Millennium Edition (ME).

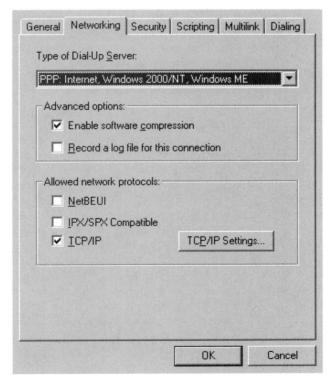

Figure 2.1 The Networking tab on the Windows Dial-Up Networking Properties dialog box.

Figure 2.2 illustrates the positioning of the Security tab on the previously mentioned dialog box to the foreground of the box. Note that under the Authentication area of the dialog box you can enter a user name, password and domain. If you look at the lower portion of the dialog box Security tab shown in Figure 2.2, you will note the label 'Advanced security options.' Under that label you will note three entries. By default each of those entries are not checked and are thus disabled. Selecting or checking the box to the left of each entry enables the entry. In Figure 2.2 this author checked the second entry to specify that only encrypted passwords can be sent to or accepted by the computer. This action provides an additional level of security for the connection. That is, when the client initiates a session and the box to the left of the label 'Require encrypted password' was checked, this results

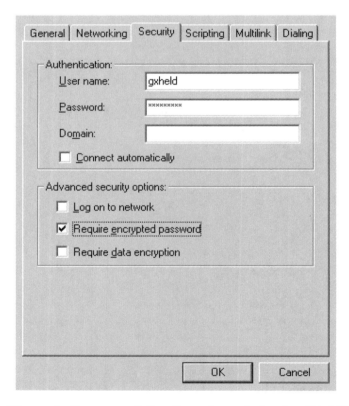

Figure 2.2 Through the use of the Security tab in the Dial-Up Networking Properties dialog box, you can define the user name and password for authentication as well as specify the use of encrypted passwords.

in the computer encrypting the password. Obviously the called computer must support encrypted passwords for this option to take effect. Since we just looked at how we can configure Windows Dial-Up Networking to use an encrypted password, let's turn our attention to the scheme by which this occurs. That scheme is referred to as the Challenge-Handshake Authentication Protocol.

2.1.2 Challenge-Handshake Authentication Protocol

The Challenge-Handshake Authentication Protocol (CHAP) represents a more secure authentication scheme than PAP. Under CHAP a three-way hand-shaking process occurs. In this section we will examine the CHAP three-way handshake as well as discuss two variations of CHAP developed by Microsoft.

2.1.2.1 Operation

Under CHAP, when a client establishes a connection to a server, the server responds to the client with a 'challenge' message as illustrated at the top of Figure 2.3. The client uses a one-way hash function to generate a response to the server. Here the term 'one-way hash' function references a function that is easy to compute but whose inverse is mathematically difficult to perform. Typically, a one-way hash function operates upon a variable length message and provides a fixed length value. For illustrative purposes, assume our one-way hash function is simply the remainder of the sum of the ASCII values of the characters in a message block divided by 256. Then, assume the message we received was:

ABADABADO

The message and the ASCII values for each character are shown below:

Message characters	A	B	A	D	A	B	A	D	O
ASCII values	65	66	65	68	65	66	65	69	79

Then, the hash value of our 'important' message becomes:

$$= \frac{R\sum(65 + 66 + 65 + 68 + 65 + 66 + 65 + 68 + 79)}{256}$$

$$= \frac{R\sum(607)}{256} = 95$$

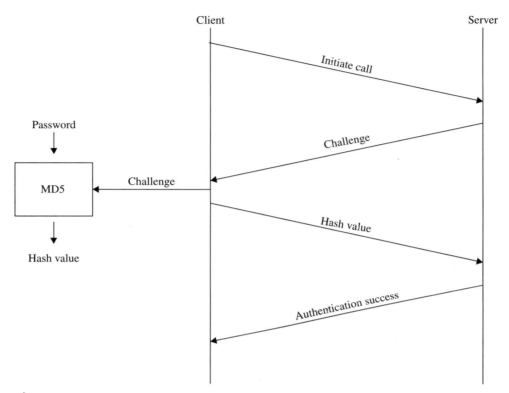

Figure 2.3 The CHAP authentication process.

Although we will discuss hash functions in more detail in the second section in this chapter, for now we only need to know that the function operates on a variable length message to produce a fixed length value that is non-reversible.

Returning to our discussion of CHAP, the client returns the value computed by applying the one-way hash function to the challenge message. This action is shown in the third line in Figure 2.3. In our example the client transmits the value 95 to the server. The server uses the same one-way hash function to compute its own local value, which is then compared to the received response. If the values match, the authentication process is acknowledged (as indicated by the fourth line in Figure 2.3). Otherwise, the connection is either terminated or the process is repeated a set number of times.

CHAP is defined in RFC 1334. Under CHAP the transmission of a user's password in the clear is avoided and the challenge normally consists of a session identifier (session ID) and an arbitrary challenge string. Message

Digest 5 (MD5), which is described in the second section in this chapter, is used as the one-way hashing algorithm to return the user name and an encryption of the challenge, session ID and the client's password, with the user name sent in the clear.

2.1.2.2 Advantages of Use

Because the user's password is not transmitted in the clear, CHAP represents an improvement over PAP. In addition, because CHAP uses a changing session ID and arbitrary challenge string for each authentication attempt, its use precludes the potential success of a replay attack. Because CHAP periodically transmits challenges to the remote client during an online session, it also protects against the possibility of a third party impersonating the remote client, a situation referred to as a 'man-in-the-middle' attack.

2.1.2.3 Derivatives

In addition to PAP and CHAP two derivatives of CHAP commonly in use are Microsoft's versions of the latter. Referred to as Microsoft Challenge-Handshake Authentication Protocol (MS-CHAP) and MS-CHAPv2, both versions slightly improve upon CHAP.

Microsoft Challenge-Handshake Authentication Protocol (MS-CHAP) represents a modified version of CHAP. Similar to CHAP, once a remote client establishes a connection to a server, the latter responds with a challenge. The challenge consists of a session ID and an arbitrary challenge string. The remote client responds with the user name and an earlier version of the Message Digest, an MD4 hash of the challenge string, the session ID, and an MD4 hash of the client's password. The use of a hash of the client's password provides an additional level of security as it permits hashed passwords to be stored on the server. In contrast, under CHAP clear text passwords are stored on the server.

A second difference between CHAP and MS-CHAP concerns the scope of protocol error codes. Under MS-CHAP protocol error codes were expanded. For example, under MS-CHAP a password expired code was added. A second version of MS-CHAP, referred to as MS-CHAPv2, offers a few additional security features beyond MS-CHAP. Under MS-CHAPv2 authentication is bi-directional, with server authentication now supported. In addition, the server will collect authentication data and can validate it against its own database or a central authentication database. This allows an organization to use a Remote Authentication Dial-In User Service (RADIUS) server or a similar centralized facility.

2.1.3 Extensible Authentication Protocol – Transport Level Security

The Extensible Authentication Protocol (EAP) represents an extension to the Point-to-Point Protocol (PPP). EAP, as its name implies, extends support for additional authentication methods. By using EAP you can consider the use of token cards, one-time passwords, public key authentication, digital certificates, and other techniques that may even be in the development stage at the present time.

While EAP was originally developed for use with PPP, it has been adopted for use with the IEEE 802.1x standard. That standard, referred to as Network Port Authentication, results in network switch ports as well as wireless access points controlling access to wired and wireless networks. Thus, EAP provides the mechanism to authenticate users who require access to wired networks or via an access point onto a wired network.

EAP-Transport Level Security (TLS) represents an extension to EAP that supports mutual authentication. Thus, under EAP-TLS, both the remote client and server must provide proof of their identities.

2.1.3.1 Limitations

Although EAP has a considerable level of support, it also has a number weaknesses or limitations. Those weaknesses include a lack of protection concerning user identity, no standardization mechanism for key exchange, a lack of support for fragmentation and reassembly and the inability to perform a fast reconnect. These limitations were in the process of being addressed by an Internet Draft entitled Protected EAP Protocol (PEAP), which wraps the EAP protocol within TLS. Under PEAP any EAP method is provided with built-in support for key exchange, session resumption and fragmentation and reassembly.

2.1.4 Token authentication

One of the more popular authentication methods beyond that of the user-ID/password pair is obtained through the use of a token. The token is commonly generated through electronic circuitry built into a credit card-sized smart card. The card typically includes a six or eight digit display and generates a pseudo-random number sequence every minute.

2.1.4.1 Operation

A remote client using a token generator is typically routed to a remote access server (RAS) for authentication purposes. The user will be prompted to enter their username, the token value displayed on their smart card and a PIN number. The PIN number is used to prevent the use of a smart card whose loss has not been reported. As a mechanism to enhance the security associated with the use of token authentication, the client will hash the username, PIN and token with predefined information known only to itself and the server. For additional security only a portion of the hashed string is transmitted to the server for authentication.

In addition to smart cards, there are other types of token generators. For example, another common token generator is produced in the form of a key fob.

2.1.4.2 Server Operation

Regardless of the way in which the token generators are produced, they all work similar to the manner previously described. Perhaps the main difference between token generators resides in the type of server used to authenticate the client's token. Some token software can be installed on RADIUS servers, while other software modules may be available for use on proprietary devices. Now that we have an appreciation of the diverse series of authentication methods that can be employed with VPNs, let's turn our attention to the second security pillar, encryption.

2.2 Encryption

In this section we will turn our attention to the wonderful world of communications encryption. First we will briefly examine how communications encryption operates and then discuss private and public key methods. Then we will turn our attention to digital certificates and the public key infrastructure (PKI), hashes, and message digests.

2.2.1 General method of operation

The function of any encryption technique is to hide the contents of a message. The contents of the message prior to the application of an encryption technique is referred to as plaintext or cleartext. In contrast, the contents of the resulting encrypted message is referred to as enciphered text or ciphertext.

In a data communications environment the most common way to encrypt data is the generation of a pseudo-random string that is modulo-2 added to the binary value of the data. The pseudo-random string is initiated with a key and the length of the key in bits as well as the encryption algorithm are major factors regarding the levels of security the algorithm provides.

Encryption at transmitter

Plaintext data to be encrypted	1000001
Pseudo-random sequence	<u>0101101</u>
Modulo-2 addition result	
Provides encrypted data	1101100

Decryption at receiver

Received encrypted data	1101100
Pseudo-random sequence	<u>0101101</u>
Modulo-2 addition result	
Restores plaintext	1000001

Figure 2.4 Encryption and decryption occurs through modulo-2 operations.

The top portion of Figure 2.4 illustrates the encryption of the ASCII uppercase A whose binary value is 1000001 or decimal 65. Note that the pseudo-random sequence used to perform the modulo-2 addition with the data results from using a specific key value with the encryption algorithm. The resulting encrypted data is transmitted. Theoretically, an unauthorized third party who intercepts the encrypted data and who knows the encryption algorithm cannot decrypt the data within a reasonable period of time. This is due to a lack of knowledge concerning the encryption key used to initiate the pseudo-random sequence. In general, the longer the key, the less vulnerable the encryption is to a brute force attack method where all key combinations are tried as a mechanism to locate the key used. For example, a 24-bit key provides 2^{24} possible key sequences, while a 36-bit key provides 2^{36} possible key sequences, and so on.

Until the 1990s perhaps the most popular encryption algorithm in use was the Data Encryption Standard (DES). DES was originally developed by IBM and submitted to the National Institute of Standards and Technology (NIST), originally known as the National Bureau of Standards (NBS). DES represents a 56-bit key which, when standardized in 1997, provided a relatively secure barrier to a brute force attack. This barrier was due to the fact that a brute force attack in which every combination of 56 1's and 0's was tried would require over a quadrillion possible keys. Using a computer capable of trying one million keys per second would require a period of time exceeding a thousand years to cycle through each possible key. Even if an unauthorized third party located the correct key halfway through the series of combinations, the information decrypted after 500 years would more than likely be of little value. Thus, in 1997 DES was considered to be secure from a brute force attack.

If we fast-forward to the present it is obvious from advertisements for 2 GHz personal computers that for a few thousand dollars you can obtain several orders of magnitude of computational capability beyond what was obtainable approximately a quarter of a century ago. In fact, over the past few years several well-publicized events with respect to the cracking of DES have occurred. In 1997 DES was cracked in a 96-day period, while in 1999 the encryption key was determined in less than 24 hours. The vulnerability of DES resulted in the development of triple DES, in which three separate keys are used. This represented an interim measure as the NIST has recently approved the Advanced Encryption Standard (AES) that can be used with 128-bit, 192-bit and 256-bit keys. However, because each extra bit position doubles the number of key combinations, triple DES (3DES) should provide many additional years of secure communications.

Returning to Figure 2.4, the encrypted transmitted data then becomes the encrypted received data. Assuming the same key is used at the receiver, the same pseudo-random sequence of bits will be generated. This type of key use in which the same key is used for encryption and decryption is referred to as a symmetrical key. Another Modulo-2 addition operation is then performed, this time between the received encrypted data and the pseudo-random bit sequence, resulting in the restoration of the plaintext. The modulo-2 operation represents an exclusive OR (XOR) operation that is relatively easy for computers to perform.

2.2.2 Private versus public key systems

DES represents a symmetric, private key encryption system. In a private key system the key remains a secret and it is only known to the parties that exchange secret messages. This results in two types of problems. First, if additional parties want to exchange secret messages, the distribution of keys can become a problem. In addition, as the number of parties exchanging secret messages increases, the number of additional keys increases. Collectively this results in a need for a key management and distribution system. Because of these problems most DES applications occur on point-to-point communications between fixed locations, such as encryption on a leased line linking a branch office to a regional center or the regional center to the corporate head-quarters. In a modern communications environment where remote clients can access literally tens of thousands of Web sites, DES and similar private key encryption methods would be impractical from a key distribution perspective. Fortunately, a second category of encryption solves the key distribution problem. That encryption method is referred to as public key encryption.

2.2.3 Public key encryption

A public key encryption system is based upon the use of a pair of separate encryption keys. The concept behind public key encryption dates back to the work of Martin Hellman and Whitfield Diffie, with the latter publishing a paper in 1976 on the use of separate keys. Diffie's paper was read by Ronald Rivest at MIT. Rivest worked with two associates at MIT, Adi Shamir and Leonard Adleman, to develop the mathematical algorithms that would make Diffie's paper a reality. Not only did Rivest, Shamir and Adleman succeed in developing the mathematical algorithm necessary for a public key system but their effort was commercialized by the company that bears the initials of the three pioneers – RSA.

2.2.3.1 Overview

A public key system is based upon the use of two keys. The public key, as its name implies, is made known to everyone. The second key, referred to as the secret or private key, is known only to the owner.

The mathematical algorithm used in a public key system permits any individual with a person's public key to encrypt cleartext data and transmit it to the owner of that key. The owner uses their private or secret key to decrypt the encrypted data. In the reverse direction cleartext encrypted with the owner's private key can only be decrypted with the recipient's public key.

2.2.3.2 Authentication Property

Prior to turning our attention to the mathematics behind public key encryption, a few words are in order concerning a key property, no pun intended, of this encryption technique. That property is the fact that the use of a public key encryption system can also provide authentication. Because a pair of keys, public and private, are used to encrypt and decrypt a message, this means that if a person can successfully decrypt a message with the originator's public key, then the message must have been encrypted with the originator's private key. Since the private key is known only to the originator, this verifies the identity of the originator. In fact, encryption performed through the use of a private key forms the basis for digital signatures that are discussed later in this chapter.

2.2.4 The RSA algorithm

There are several mathematical algorithms that can be used in a public key system. By far the most popular algorithm in use is the RSA algorithm, developed by Rivest, Shamir and Adleman. The RSA algorithm is based upon the use of very large prime numbers and their mathematical manipulation. Thus, prior to describing how encryption and decryption occurs, a brief digression to review both prime numbers and their mathematical manipulation will be beneficial, so let's do so.

2.2.4.1 Prime Numbers

From high school math we probably remember that a prime number is a number that has only two factors, 1 and itself. For example, 1, 3, 5 and 7 all represent prime numbers as they can be divided by 1 and themselves. In comparison, 4, 9 and 12 are not prime numbers as they are evenly divisible by at least one other number than 1 and by themselves. For example, 4 is divisible by 1, 2, and 4, while 9 is divisible by 1, 3 and 9.

Another prime number concept that is important to note is relatively prime. Two numbers are considered to be relatively prime if they share no common factors other than 1. For example, 18 and 25 are relatively prime although neither number is a prime. The factors of 18 are 1, 3 and 9, while the factors of 25 are 1 and 5. Thus, the only common factor of both numbers is 1, resulting in them being relatively prime to one another.

2.2.4.2 Modular Mathematics

Earlier in this chapter when we examined the manner by which encryption and decryption occurs, we noted the use of modulo-2 operations. When the RSA algorithm is used in public key encryption and decryption, modular mathematical operations are used with prime numbers. When working with modular mathematical operations it is convenient to think of the modulus as a maximum, which, when reached, requires you to start over again. Thus, some examples of modular mathematical operations include:

$$12 \bmod 9 = 3$$

$$21 \bmod 9 = 3$$

$$30 \bmod 9 = 3$$

2.2.4.3 Exponential Operations

Because the RSA algorithm is based upon the use of exponential values of prime numbers, another area of mathematics we need to review is exponential operations. In doing so there are two exponential identities we need to consider. The first exponential identity states that a number raised to one power multiplied by the same number raised to a different power is equal to the number raised to the sum of the two powers. Mathematically, we can denote this as:

$$X^a * X^b = X^{(a+b)}$$

For example, assume a value of 3 for X and 2 and 3 for the exponents a and b, respectively, Then,

$$3^2 * 3^3 = X^{(2+3)}$$

or

$$3^2 * 3^3 = 3^5$$

or

$$9 * 81 = 729$$

Regardless of the numbers you use for x, a and b, the left and right sides of the prior exponential identity will remain equal.

A second exponential identity worth noting concerns the raising of a number to a power and then raising the result to a second power. That is $(X^a)^b$. The second exponential identity states that this is equal to raising the number to the product of the two exponents. Mathematically, this exponential relationship can be expressed as follows:

$$(X^a)^b = X^{a*b}$$

For example, again using a value of 3 for X and 2 and 3 for exponents a and b, we obtain:

$$(3^2)^3 = 3^{(2*3)}$$

or

$$9^3 = 3^6$$

or

$$729 = 729$$

2.2.4.4 Algorithm Operation

Now that we have an appreciation of the basic mathematical operations, let's turn our attention to the RSA public key algorithm. The steps in the algorithm are as follows:

1. Select two large primes referred to as p and q.
2. Take the product of the two primes ($n = p * q$) and use n as the modulus.
3. Select a number, e, which is less than n and relatively prime to $(p - 1)(q - 1)$. From our prior discussion this means that e and $(p - 1)(q - 1)$ have no common factors other than 1.
4. Find the inverse d mod $(p - 1)(q - 1)$ such that $e * d = 1 \bmod (p - 1)(q - 1)$.

Here e and d represent the public and private exponents. The public key represents the pair of values (n, e) while the private key is d. The primes p and q are kept secret since they define the keys.

To encrypt a message, m, a user would create ciphertext, c, using the public key pair n and e as follows:

$$c = m^e \bmod n$$

To decrypt the received enciphered data the receiver uses their private key d as follows:

$$m = c^d \bmod n$$

Note that the RSA algorithm uses two modulus. One modulus is used to create the keys while a second is used for encryption and decryption.

To illustrate the manner by which the RSA algorithm operates, we will use two small prime numbers to facilitate computations instead of large primes. Thus, for step 1 of the algorithm let's assume we selected $p = 5$ and $q = 11$. Under step 2 of the algorithm we multiply the primes to obtain the modulus. Thus, $5 * 11$ or 55 represents n.

The third step in the algorithm is to select a number e that is less than n (55) and relatively prime to $(p - 1)(q - 1)$. This requires us to find a value for e that has no common factors with $(p - 1)(q - 1)$. Since $(p - 1)(q - 1)$ is equivalent to $(5 - 1)(11 - 1)$ or 40, we need to find a value for e that has no common factors other than 1 with 40. Here one common value for e would be 3.

The fourth and last step is to find the inverse d mod $(p - 1)(q - 1)$ such that:

$$e * d = 1 \bmod(p - 1)(q - 1) \text{ or}$$

$$3 * d = 1 \bmod 40$$

As a refresher, the inverse of any whole number is simply 1 divided by the number, which results in multiplying a number by its inverse, yielding a value of unity. In modular mathematics an inverse represents two whole numbers that when multiplied together result in a value of 1. Returning to our example, we need to determine a value for d such that when multiplied by 3 yields 1 mod 40.

We could cycle through a sequence of numbers to determine d but this would be time-consuming, especially when large prime numbers are used. As an alternative, we could use the extended Euclidean algorithm, which consists of a series of multiplications and subtractions to determine the inverse value.

To illustrate the basic operation of the extended Euclidean algorithm to solve $3 * d = 1$ mod 40, we would create two columns of numbers as follows:

$$
\begin{array}{cc}
40 & 40 \\
3 & 1
\end{array}
$$

Using the algorithm, we perform operations on the left column to find values to modify the right column. First, we will multiply the second row by a number such that the result is as close to the value of the number in the first column without exceeding the number. For example, let's use 13. Then, we obtain:

$$
\begin{array}{cc}
40 & 40 \\
3 * 13 = 39 & 1 * 13 = 13
\end{array}
$$

Next, we subtract the second row from the first to obtain:

$$
\begin{array}{cc}
40 & 40 \\
3 * 13 = 39 & 1 * 13 = 13 \\
40 - 39 = 1 & 40 - 13 = 27
\end{array}
$$

We then bump up each row by removing the top row to obtain:

$$
\begin{array}{cc}
39 & 13 \\
1 & 27
\end{array}
$$

Next, we would attempt the same multiplication operation; however, when the result is 1 in the left column, we would stop and take the number in the right column as the inverse. Thus, based upon the preceding series of operations, the value of the inverse (d) was determined to be 27. Now that we determined the values for d, e and n, we would make n and e public and keep d as well as p and q private or secret. Thus, to encrypt a message m, the cyphertext (c) is computed as follows:

$$c = m^e \bmod n$$

Let's assume we want to encrypt the message ENDGAME. For simplicity, let's focus our attention upon the encryption and decryption of the first letter in the message, the character E, whose numeric value is 5. Using the previously noted formula for creating ciphertext we obtain:

$$c = 5^3 \bmod 55$$

$$c = 125 \bmod 55$$

$$c = 15$$

Thus, the fifth character in the alphabet whose numeric value is 5 would be encoded with a value of 15 when encryption is based upon the previously selected prime numbers.

To decrypt the received message of the previously encrypted character the private key is used as follows:

$$m = c^d \bmod n$$

Thus,

$$m = 15^{27} \bmod 55$$

Because any number to a large power represents a serious effort beyond the computational capability of most calculators, we can simplify the computations by remembering the following relationship:

$$X^a * X^b = X^{a+b}$$

Thus, we can note the following relationships:

$$15^1 = 15 \ \times 1 \quad = 15 = 15 \bmod 55$$

$$15^2 = 15 \ \times 15 \ = 225 = 5 \bmod 55$$

$$15^4 = 15^2 \times 15^2 = (5 * 5) \bmod 55 = 25 \bmod 55$$

$$15^8 = 15^4 \times 15^4 = (25 * 25) \bmod 55 = 20 \bmod 55$$

$$15^{16} = 15^8 \times 15^8 = (20 * 20) \bmod 55 = 15 \bmod 55$$

From the above relationships we can compute 15^{27} as:

$$15^{16} * 15^8 * 15^2 * 15^1 \text{ or}$$

$$(15 * 20 * 5 * 15) \bmod 55 \text{ or}$$

$$22500 \bmod 55 \text{ or } 5$$

Thus, the cleartext value of the fifth letter in the alphabet (E) is obtained, which recovers the plaintext from the use of the private key.

2.2.4.5 Limitations

Based upon the previous example of public key encryption note that the computation of each encrypted character occurs via an exponential and a modulus operation. These two operations are significantly more processor intensive to perform than operations used to encode data when a private key system is used to generate a pseudo-random string that is XORed with plaintext. Thus, the mathematical operations represent a key limitation of public key encryption. That is, it is highly processor intensive. In fact, public key encryption and decryption are approximately a thousand times slower than secret key operations, such as DES. This means that the encryption of a message that takes a few seconds when DES is used can require a half-hour or more when using a public key encryption. While this time constraint can be expected to put a damper on the possible use of public key encryption for real-time communications, the use of a combination of public and private key systems can overcome timing problems as well as provide authentication and solve the key distribution problem associated with the use of private key encryption. To accomplish the preceding, public key encryption is first used to exchange private keys. Then, once the private keys have been exchanged, they are used for encrypting and decrypting data. Now that we have an appreciation of the basic operation of public and private key systems and their constraints, let's turn our attention to the role of digital certificates and digital signatures.

2.2.5 Digital certificates

A digital certificate physically represents an attachment to an electronic message. This attachment contains information that verifies the identity of the person sending the message. In this section we will discuss the role of the certificate authority in issuing a digital certificate as well as the use and contents of the certificate.

2.2.5.1 The Certificate Authority

Digital certificates are issued by a certificate authority (CA). The certificate authority represents a trusted third party that guarantees that the person that has the certificate or, more accurately, the person that placed the certificate on a computer, is who they claim to be. The digital certificate authority (DCA) issues a unique digital certificate to each applicant. That certificate contains the applicant's public key and other information, which we will shortly discuss. The digital certificate is signed using the CA's private key. Because the CA's public key is readily accessible via the Internet or was previously distributed, the validity of the certificate can be established, permitting the certificate holder to be authenticated. This authentication is accomplished by the recipient of an encrypted message using the certificate authorities' public key to decode the digital certificate attached to the message. Because the sender's public key is contained within the certificate, the recipient can respond with an encrypted reply.

Figure 2.5 illustrates the triangular relationship between a certificate authority, the message originator who was issued a certificate, and a message recipient. Note that the message recipient will first receive a copy of the certificate from the applicant that contains information about the certificate authority. Using this information the message recipient will retrieve the certificate authorities' public key. Because the digital certificate was signed with the certificate authorities' private key, the recipient can use the CA's public key to verify or authenticate the certificate. Since the certificate contains a copy of the message originator's (applicant's) public key, the recipient will then use that key to respond to the originator with encrypted messages that the originator decrypts using its private key. Now that we have an idea of how digital certificates can be used, let's look at their contents.

2.2.5.2 Certificate Contents

Similar to most aspects of data communications, there is an international standard which governs the contents of digital certificates. That standard is the International Standards Organization (ISO) X.509 standard. As we turn our attention to the contents of a digital certificate, we will literally 'kill two birds with one series of screen images' and discuss the contents of certificates as well as note where they can be viewed in a Microsoft Windows environment.

If you are using Microsoft's Internet Explorer and click on the Tools menu you will locate Internet Options on the displayed menu. Selecting that entry results in the display of a dialog box with six tabs. Figure 2.6 illustrates the Internet

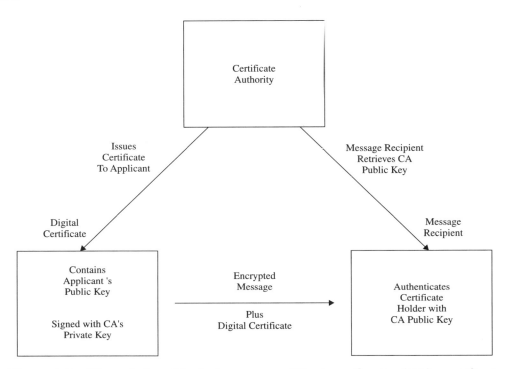

Figure 2.5   The relationship between a certificate authority (CA), certificate holder, and message recipient.

Options dialog box with its Content tab displayed in the foreground. Note that the middle section of the tab display concerns the use of certificates. The first button, which is labeled 'Certificates,' results in a listing of digital certificates known to Internet Explorer at this particular point in time. The second button, which is labeled 'Publishers,' lists those organizations you previously designated as trustworthy. Once so designated, Windows applications can install and use software from listed vendors without prompting for permission.

If you click on the button labeled Certificate, this will result in the display of a dialog box labeled 'Certificate Manager,' which has three tabs and a pull-down selection menu that governs the display of certificates by their purpose. Figure 2.7 illustrates the Certificate Manager dialog box with its Intermediate Certification Authorities tab shown displayed in the foreground. Note the pull-down menu associated with the 'Intended' purpose label is by default set to 'all.' Other selections available include Client Authentication, Secure Email, and Advanced Purposes.

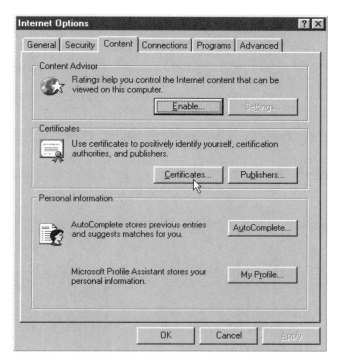

Figure 2.6   The Content tab in Internet Explorer's Internet Options dialog box provides users with the ability to view and edit digital certificates.

The primary use of Certificate Manager is to facilitate the display of a list of certificates installed on a computer. The list displayed will be based on the tab and intended purpose selection. In addition, through the use of Certificate Manager you can obtain the ability to view the contents of individual certificates as well as enable or disable the use of the certificate for certain operations.

Returning to Figure 2.7, we selected the Microsoft Windows entry. If you look at the lower left portion of the display you will note that the intended purpose of the selected digital certificate is displayed. In this example, the purpose is to verify Windows hardware drivers. If you click on the button labeled 'View' shown in the lower right portion of Figure 2.7 you can obtain both general and specific information concerning a selected certificate as well as its certification path. Here the term certification path represents a chain of related certificates. Because we previously selected the certificate issued to Microsoft Windows, let's view that certificate.

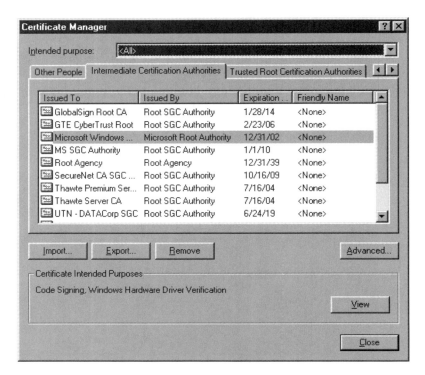

Figure 2.7 The Certificate Manager dialog box provides Internet Explorer users with the ability to view and manage digital certificates.

2.2.5.3 Viewing Certificate Details

Figure 2.8 illustrates the certificate dialog box displayed when you either click on the View button or double-click on a specific certificate entry. Note the Certificate dialog box has three tabs, with the tab labeled 'General' shown in the foreground by default. That tab provides general information about the certificate to include its intended use, who it was issued to and who issued the certificate, as well as its validity period.

For the certificate we are viewing we can note from Figure 2.8 that the General tab informs us that its use is to verify the Windows hardware driver, that it protects software from tampering after publication and ensures that software came from the software publisher. We can also note that the certificate was issued by the Microsoft Root Authority to Microsoft Windows Hardware Compatibility and the certificate is valid from October 1, 1997 through December 31, 2002.

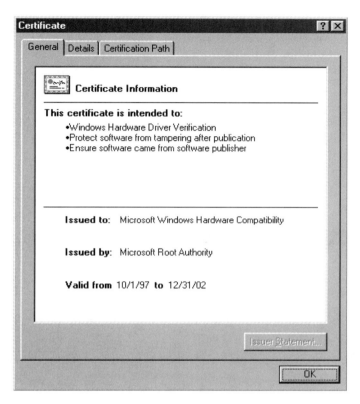

Figure 2.8 The General tab in the certificate dialog box provides information concerning the use of the certificate, the issuing authority and the period for which the certificate is valid.

As a brief digression, there are three types of certificate authorities: self-signed, subordinate and root.

- In a self-signed CA the public key in the certificate and the key used to verify the certificate are the same. Some self-signed CAs are root CAs, as we will shortly note.
- In subordinate CA, the public key in the certificate and the key used to verify the certificate differ. When one CA issues a certificate to another CA, the process is referred to as cross-certification.
- The root CA is a special class of certificate authority. The root CA is located at the top of a certification hierarchy and is unconditionally trusted by its clients. Because there is no higher certifying authority, the root CA

must sign its own certificate. While all root CAs can be considered to be self-signed CAs, the opposite is not true.

Returning to the certificate we previously selected, for additional information let's examine the use of the Details tab.

The display of the Details tab results in the ability to observe specific information concerning the selected certificate. Figures 2.9 and 2.10 illustrate all of the fields in the selected certificate. Figure 2.9 shows the values for the first eight fields in the certificate. By scrolling down, we are able to view the values of the remaining four fields as shown in Figure 2.10. Note from Figure 2.9 that by clicking on a particular field you can display specific

Figure 2.9 Selecting the Details tab on the certificate dialog box provides the user with the ability to examine the contents of the fields of the selected authority.

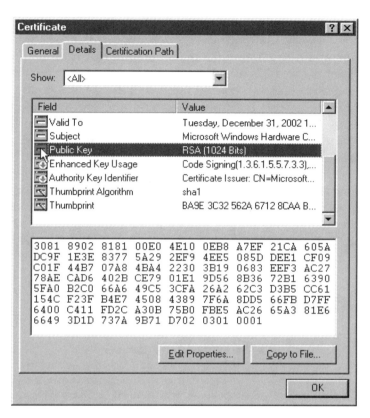

Figure 2.10 By clicking on a certificate field you can display specific information contained in the field. In this example we are viewing the contents of the public key.

information concerning the contents of the field. For example, in Figure 2.10 we can observe the contents of the public key contained in the certificate.

The icons to the left in both figures indicate if the field represents an X.509 Version 1 field, a non-critical X.509 Version 3 extension (downward arrow), a critical X.509 Version 3 extension (triangle superimposed on the icon), or an editable property associated with the certificate (pencil superimposed on the icon).

In examining the fields shown in Figures 2.9 and 2.10 you can obtain an idea of the contents of a certificate. Because Figure 2.10 shows both the selection of a field and the scrolling to the last field, we will use the fields in both Figures 2.9 and 2.10 to discuss the contents of a certificate. Beginning with

Figure 2.9, the first field is the Version field. This field indicates the X.509 version number. To facilitate our discussion concerning the information in a digital certificate, Table 2.1 provides a description of each of the certificate fields shown in Figures 2.9 and 2.10. In addition, in the lower portion of Table 2.1 a description of two common optional fields will be found.

In examining the entries in Table 2.1 in conjunction with the field values of the certificate shown in Figures 2.9 and 2.10, we can obtain a significant amount of information about this digital certificate. For example, we can note the serial number of the certificate as well as the fact that the RSA algorithm was used with the MD5 hash to generate the issuer's signature. The fourth field indicates the issuer of the certificate, which is the Microsoft Root Authority.

Continuing our tour through the fields shown in Figure 2.9, the next two fields indicate that the certificate is valid from October 1, 1997 through December 31, 2002. The following field, which is Subject, indicates that the owner of the certificate is Microsoft Windows Hardware. The last field shown

TABLE 2.1 Digital Certificate Fields

Certificate Field	Description
Version	The X.509 version number
Serial number	A unique number the issuing certificate authority assigns to the certificate for administrative purposes.
Signature algorithm	The hash algorithm used by the certificate authority to sign the certificate.
Issuer	The name of the organization that issued the certificate.
Valid from	The beginning date for the period in which the certificate is valid.
Valid to	The final date for the period in which the certificate is valid.
Subject	A name which uniquely identifies the owner (individual or certification authority) to whom the certificate was issued.
Public key	The owner's public key type and length.
Thumbprint algorithm	The hash algorithm used to generate a digest of data. That digest for thumbprint is used for digital signatures.
Thumbprint	The digest of the certificate data.
Common Optional Fields	
Friendly name	A friendly, or common, name for the name listed in the Subject field.
Enhanced key usage	The purpose for which the certificate was issued.

in Figure 2.9 prior to the scrolling shown in the next figure is the Public Key field. In Figure 2.9 we can note that the 128-bit RSA public key was used.

If we turn to Figure 2.10, we can note that in addition to observing the 128-bit public key this illustration also shows a continuation of the fields in the selected certificate. If you turn to the description to the right of the Enhanced Key Usage field you will note a series of numbers separated by decimal points followed by a series of three dots, the latter indicating additional information follows. The first sequence of digits is a global identifier that indicates Code signing. A second sequence of numerics would be observed if you moved the highlighted bar over the field. That sequence of numerics would indicate the position in the global identifier for Windows Hardware Driver Verification.

The Authority Key Identifier field indicates information about the organization that issued the certificate. If you look at the value column you can see some information about the authority key identifier without placing the highlighted bar over the field. In Figure 2.10 we can note that the certificate issuer is Microsoft. As previously mentioned, the thumbprint algorithm is the hash algorithm used to generate a digest of data that Microsoft references as a thumbprint. This thumbprint is used for digital signatures and the actual thumbprint is the last field shown in Figure 2.10. Thus, in continuing our discussion of encryption and digital certificates we would be remiss if we did not discuss digital signatures – so let's do so.

2.2.6 Hashing and digital signatures

In this section we will look at the manner in which a digital signature is formed. Because a hash or message digest algorithm must be applied to a message to generate a digital signature, we will first turn our attention to the former prior to discussing the latter.

2.2.6.1 Hashing and Hash Algorithms

If we return to Figure 2.10 for a moment we will note that the thumbprint is shown as 'SHA1.' That entry represents one of two widely used hash functions or algorithms, the other being MD5.

The Secure Hash Algorithm 1 (SHA1) was developed by the National Institute of Standards and Technology and the National Security Agency (NSA) as a mechanism to create digital signatures. SHA1 was based upon Message Digest 5 (MD5), which was developed by Ron Rivest, one of the founders of RSA. As we noted earlier in this chapter, hashing involves the conversion of a message into a shorter, fixed-length value. In effect, a hash function or algorithm can be considered to generate a special type of compressed version of a file. However, unlike conventional compression, the hash result should not be reversible.

2.2.6.2 Hash Properties

If we consider a message to represent a string of characters, then a hash function used for cryptographic purposes represents an algorithm that must have certain properties. One property is that the algorithm generates a fixed length value that is difficult, if not impossible, to invert. This means that the hash function should be a one-way function. Mathematically, if H is a hash function and $H(x) = h$, then we can say it is a one-way function if it is computational infeasible to determine x.

A second property of a cryptographic hash function is that it needs to be collision free. Here the term collision free means that it should be computationally infeasible to have two messages, x and y, such that $H(x) = H(y)$. If we briefly turn our attention to the elementary function described earlier in this chapter, it is obvious that that function is not collision free. Thus, simply adding the ASCII value of each character in a message and dividing by a fixed quantity and using the remainder might be useful as a checksum but would not represent a cryptographic hash function.

A third property of a hash function is that it should be relatively easy to compute. Two additional properties for a cryptographic hash function include supporting operations on a message of any length and generating a fixed length output. Thus, the basic properties of a cryptographic hash function are:

- relatively easy to compute;
- one-way computation;
- collision free;
- operates on a variable length message;
- generates fixed length output.

2.2.6.3 Message Digest Algorithms

The original message digest (MD) as well as subsequent digests were developed by Ron Rivest. The original message digest in the series was proprietary and never released to the public. The second message digest, MD2, was developed by Rivest in 1989. Under MD2 a message is first padded so that its length in bytes is divisible by 16. A 16-byte checksum is then appended to the message and the hash value is computed on the resulting message. The hash value is computed by processing the message in 16-byte chunks, each generating an intermediate result. Although MD2 has not been compromised, it was determined that collisions could be constructed if the calculation for the checksum was omitted, resulting in Rivest developing two additional message digests.

MD4 and MD5 were developed in 1990 and 1991, respectively. Both are similar in that they operate on 512-bit (32-word) message chunks. Under MD4 the message is padded such that its length in bits plus 64 is divisible by 512. A 64-bit binary representation of the original length of the message is concatenated to the message, after which it is processed in 512-bit blocks, with each block processed in three iterations or passes.

Several attacks on MD4 in which collisions can be generated in under a minute on a conventional PC were successfully reported. Due to this, RSA Security announced that MD4 should be considered broken.

MD5, while similar to MD4, uses an algorithm that processes each 512-bit block in four passes. The additional pass beyond that used in MD4 as well as a slightly different method of computing intermediate values results in a more secure digest. However, the computations are more processor intensive, which results in a slower but strengthened digest. MD2, MD4 and MD5 are defined in RFCs 1319, 1320 and 1321.

2.2.6.4 MD5 versus SHA1

As previously mentioned, SHA1 is based to a large extent on MD5. Both pad messages in the same manner. However, because SHA1 makes five passes over each block of data during which each word in the message digest is further manipulated, it is considered to be more secure than MD5. Similar to a comparison of MD5 to MD4, SHA1 requires more processing time than MD5. One additional difference between the two is the fact that SHA1 produces a 160-bit digest in comparison to the 128-bit digest generated by the MD4 and MD5 algorithms.

2.2.6.5 Creating a Digital Signature

The application of a hash function to a message generates a thumbprint or digital fingerprint. Because some well-known hash functions are known as message digests, that term is quite often used to reference both the hash function as well as the resulting thumbprint.

Once the hash of a message is created, its encryption with a private key results in what is referred to as a digital signature. In addition, this action serves as a mechanism that authenticates the message even though the hash is separate from the document.

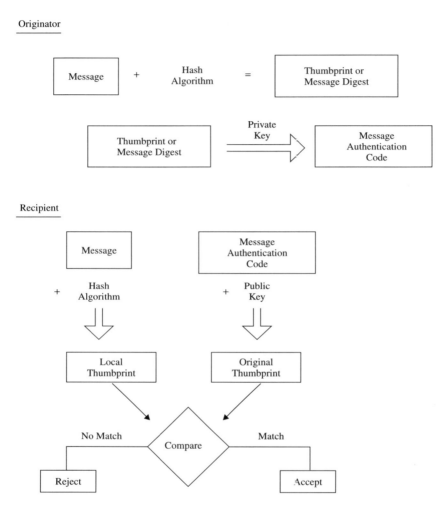

Figure 2.11 Using a digital signature.

Figure 2.11 illustrates the creation of a digital signature. The resulting hash can be considered to represent a message authentication code that has several important properties in comparison to the encryption of an entire message. First, since only the hash is encrypted, this action is much less processor intensive. Second, because the hash algorithm created a one-way function, the message recipient can apply the same algorithm to a received message. If the locally generated hash equals the transmitted hash, then the recipient knows that the message has not been altered.

chapter three

Understanding the TCP/IP Protocol Suite

Most people who visit a bookstore or scan the Internet can easily locate a variety of titles that cover the TCP/IP protocol suite in detail. While such books represent excellent references for learning the details of the protocol suite, their wealth of information makes such books difficult to read when attempting to review key areas of the TCP/IP protocol suite with respect to VPN operations. Thus, the aim of this chapter is to provide detailed information about the TCP/IP protocol suite as it relates to VPN operations in one concise location. In doing so we will first examine the manner by which LAN frames are formed to include the relationship of network and transport headers within the frame. Once this is accomplished, we will turn our attention to the network and transport layers, examining the use and settings of certain fields as they relate to VPN operations. In concluding this chapter our focus will turn to two services that can affect the operation of VPNs: network address translation and proxy services. Thus, while not intended as a comprehensive guide to the TCP/IP protocol suite, this chapter does provide a comprehensive guide to those aspects of the suite that relate to VPN operations.

3.1 Frame formation

In this section we will review the manner in which an IP datagram is formed and transported within a LAN frame. In doing so we will note the relationship between the various headers within a frame and how the positioning of those headers affects the operation of such security devices as router-programmed access lists, firewalls and VPN gateways.

Virtual Private Networking G. H. Held
© 2004 John Wiley & Sons, Ltd   ISBN: 0-470-85432-4

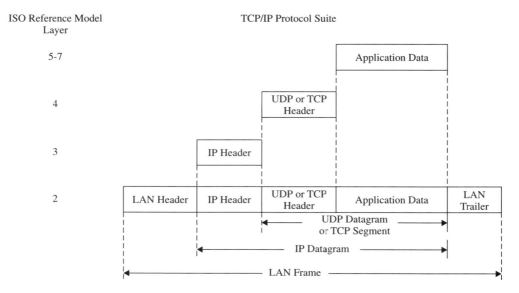

Figure 3.1   LAN frame formation.

3.1.1 Header sequencing

Figure 3.1 illustrates the relationship between the formation of data for transport by a LAN frame and the International Standards Organization (ISO) Open System Interconnection (OSI) Reference Model. In examining the formation of a LAN frame transporting an IP datagram, note that the application layer in the TCP/IP protocol suite is equivalent to layers 5 through 7 of the ISO's OSI Reference Model. As a LAN frame is formed, a TCP or UDP header is prefixed to the application data. Either header includes both a source and a destination numeric identifier that indicate the type of application data being transported. That numeric identifier is referred to as the source or destination port. In actuality, the destination port number identifies the application since a receiving device 'listens' on predefined ports for predefined applications associated with port numbers. In comparison, the source port is normally set to either a value of zero or a randomly selected value.

3.1.2 Segments and datagrams

When a TCP header is prefixed to an application data unit, the result is referred to as a TCP segment. In comparison, the prefix of a UDP header to an application data unit results in the formation of a UDP datagram. The formation of UDP datagrams and TCP segments occurs at the transport layer in the TCP/IP protocol suite, which is layer 4 in the OSI Reference Model.

The prefix of an IP header to the UDP datagram or TCP segment occurs at the network layer in the TCP/IP protocol suite. This corresponds to layer 3 in the OSI Reference Model. The IP header includes source and destination address fields. Thus, the IP header in conjunction with the TCP or UDP header includes several key fields that identify the application being transported, who originated the data, and its destination. Although TCP and UDP headers include both source and destination port fields, as briefly mentioned earlier in this chapter, the source field is normally initially set to a zero value or to a randomly selected value. As we probe deeper into the relevant fields in the headers that must be considered when configuring routers, firewalls and gateways for VPN operations, we will note some of the well-known field assignments.

Concerning the terms routers and gateways, we will consider them to be equivalent and use the terms interchangeably, although a purist might disagree. As a brief notation of an historical nature, the first type of device used to route datagrams from one network to another was called a gateway even though today such devices are marketed as routers. In fact, when you configure a workstation in a Windows environment you enter the IP address for the primary gateway and optionally for a secondary gateway. The workstation uses those IP addresses to forward datagrams whose destinations are not on the current network, resulting in the gateway routing the datagrams towards their ultimate destination. Although Microsoft maintains the term gateway in its dialog box, in effect, we are informing the workstation we are configuring where the router resides.

3.1.3 ICMP Messages

Although not shown in Figure 3.1, the transport of an Internet Control Message Protocol (ICMP) message deserves mention. ICMP messages are an integral part of the Internet Protocol that conveys error and control messages. Both routers and hosts use ICMP to transmit reports concerning received datagrams back to the originator as well as generate the well-known and frequently used echo request and echo reply, collectively and commonly referred to as Ping. An ICMP message is transported as an IP datagram. This means that the ICMP message is simply prefixed with an IP header, resulting in the encapsulation of an ICMP message within an IP datagram as shown in Figure 3.2.

Figure 3.2 An ICMP message is transported by the prefix of an IP header to the message.

3.1.4 On the LAN

Continuing our examination of the transport of an IP datagram, previously shown in Figure 3.1, note that when placed into a LAN frame the datagram is located in the Information field of the frame. The LAN header is similar to the IP header with respect to containing source and destination address fields. Those fields contain media access control (MAC) 48-bit addresses instead of 32-bit (IPv4) or 128-bit (IPv6) addresses used in the IP header.

When a router receives an IP datagram destined for a device on a LAN, it must form a layer-2 frame to deliver the datagram. This means the router needs to know the destination MAC address. To determine that address, the router will first search its internal tables to see if it previously transmitted a datagram with the same destination IP address and already learned the destination MAC address. If so, the router will use the previously learned MAC address. If the MAC address associated with the destination IP address is not known, the router will broadcast an Address Resolution Protocol (ARP) frame onto the LAN. The ARP frame in effect tells each station on the LAN to respond with its MAC address if it recognizes the IP address in the frame. Thus, the response to the ARP provides the router with the layer-2 MAC address necessary to deliver an IP datagram.

3.1.5 Dataflow control fields

When data flows to or from a LAN there are six fields that can be used as control mechanisms to enable or disable the flow of LAN frames or IP datagrams. Those fields are listed in Table 3.1 and contain some of the key metrics you need to consider when configuring routers, VPN gateways and firewalls. Such devices can be configured to enable or disable the flow of datagrams based upon a specific field value or range of values. In addition, most devices can be configured to enable or disable the flow of datagrams based upon the contents of multiple fields, such as IP destination address and TCP destination port number.

TABLE 3.1 Key IP datagram fields to consider when configuring equipment to support VPN operations

Field	Description
IP source address	Identify the originator of datagram
IP destination address	Identify the recipient of datagram
TCP source port	Identify the application program at the end of the connection
TCP destination port	Identify the application program at the end of the connection
UDP source port	Identify the application program at the end of the connection
UDP destination port	Identify the application program at the end of the connection

In addition to the six fields listed in Table 3.1, certain routers and firewalls provide the ability to alter the flow of data based upon the composition of other fields. For example, the TCP Flags field settings are used by some products as a decision criteria when appropriately configured. Thus, setting a router access list could be used to look for the setting of the TCP SYN or ACK bits, allowing packets that are part of an existing conversation to flow through the router if those bits are set. Therefore, when you attempt to establish a VPN you may need to reconfigure hardware to allow the flow of datagrams based upon certain predefined field values.

3.2 The network layer

As indicated in the prior section in this chapter, the formation of IP datagrams occurs at the network layer through the prefix of an IP header. That header has two key fields that can be used to control the flow of datagrams through various types of hardware – the source and destination IP addresses. In addition, there are several other fields within the IP header that can be used by hardware to control the flow of datagrams. Because of this we will briefly review some of the fields within the IP header in this section as well as the various types of ICMP messages, as disabling the flow of the latter under certain circumstances can adversely affect the operation of a VPN.

3.2.1 The IPv4 header

Figure 3.3 illustrates the fields within an IPv4 header. The first field is a 4-bit Version field that not only specifies the version of the IP protocol in use, it also enables the originator, recipient and gateways between the parties to agree upon the format of the datagram. As you might expect, for IP version 4 (IPv4) the value of the Version field is 0100 binary or decimal 4.

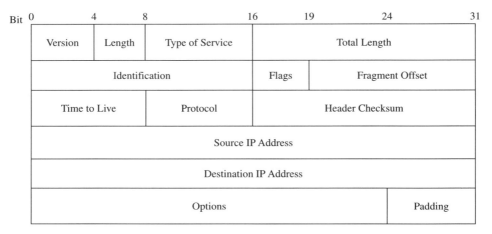

Figure 3.3   The IPv4 header.

In a VPN operating environment there are several fields in the IP header that warrant attention. Those fields include three that provide fragmentation control (Identification, Flags and Fragment Offset), the Time-to-Live field, the Protocol field, and as indicated earlier in this chapter, the source and destination IP address fields. In reviewing the previously mentioned fields, we will also briefly discuss IP addressing and the subnet mask as they not only affect the flow of data, they also form the basis for the Cisco 'wildcard mask' that is used in that vendor's access list statements to control the flow of datagrams through a router's interface.

3.2.1.1 Fragmentation Control

Three fields in the IP header control fragmentation and reassembly of datagrams. Those fields are the Identification, Flags and Fragment Offset fields.

The Identification field is a 16-bit field that contains a unique number that identifies the datagram. When a datagram is fragmented, each fragment is prefixed with an IP header that has the same number in the Identification field. At the destination the receiving device can reassemble the fragment based upon the Identification and Source IP Address field values as both uniquely identify a sequence of fragments generated by a common source.

The second field in the IP header associated with fragmentation is the Flags field. The low-ordered 2 bits in this 3-bit field control fragmentation. The first bit setting defines whether a datagram can be fragmented. Because setting this bit to 1 indicates that the datagram should not be fragmented, this bit is referred to as the 'do not fragment bit.' The low-ordered bit in the Flags field denotes whether a fragment is the last one or if more fragments follow. As you might expect, this bit is referred to as the 'more fragments' bit.

The third field associated with fragmentation is a 13-bit Fragment Offset field. The purpose of this field is to specify the offset of a fragment from the original datagram. The value of this field is specified in units of 8 bytes, with the initial value being zero, indicating the first fragment of a datagram. As we will note later in this book, when fragmenting datagrams within the context of a VPN, the method by which the VPN is formed has a direct bearing on the manner and effort required to perform the reassembly of datagrams.

3.2.1.2 Time-to-Live Field

The original intention of the Time-to-Live (TTL) field was to specify how long, in seconds, a datagram could remain in the Internet. Because time measurements are most difficult, the TTL field uses a router hop value that is decremented as the datagram flows through a device. When set to a value of zero, the datagram is discarded, providing a mechanism that prevents endless routing due to a variety of erroneous conditions. Since a hop value is used, an alternate name for this field is the hop count field.

3.2.1.3 Protocol Field

The Protocol field is an 8-bit identifier that indicates the higher level protocol transported by the IP header as a datagram. For example, the Protocol field has a value of 1 when IP transports an ICMP message, while values of 6 and 17 are used to identify the transport of TCP and UDP. In the wonderful world of VPNs the value of the Protocol field will indicate the higher layer protocol used to transport VPN data. In many network situations someone will forget to configure a router or firewall to allow the particular protocol to pass, resulting in many hours of effort by personnel attempting to identify the reason why a VPN cannot be established.

Table 3.2 lists 10 examples of assigned Internet Protocol numbers. Because an 8-bit byte is used for the protocol field, this enables up to 256 (0 through 255) protocols to be defined. Note that protocol values 117 through 254 are presently unassigned while value 255 is reserved. Of special interest are decimal values 50 and 51, which define two types of IPSec headers that can be transported as IP datagrams. In addition, decimal value 115 permits a VPN tunneling protocol, referred to as L2TP, to be transported as an IP datagram. Later in this book we will discuss the use of IPSec and L2TP.

TABLE 3.2   Examples of assigned Internet Protocol Numbers

Decimal Value	Keyword	Protocol
1	ICMP	Internet Control Message Protocol
6	TCP	Transmission Control Protocol
8	EGP	Exterior Gateway Protocol
17	UDP	User Datagram Protocol
41	Ipv6	IPv6
46	RSVP	Resource Reservation Protocol
50	ESP	Encapsulation Security Payload
54	AH	Authentication Header
58	Ipv6-ICMP	ICMP for IPv6
115	L2TP	Layer 2 Tunneling Protocol

3.2.1.4 Source and Destination IP Address Fields

Under IPv4 both source and destination IP address fields are 32 bits in length. Class A, B and C addresses, which are probably by far most commonly used in the Internet, are subdivided into network and host portions as indicated in Figure 3.4.

In examining Figure 3.4 a Class A network uses the first byte as the network addresses while the following three bytes are used to identify the host on a particular network. Although each byte can theoretically have a decimal value between 0 and 255, the first bit in the first byte is set to 0 to identify a Class A address, reducing the available networks to a maximum of 127. However, an

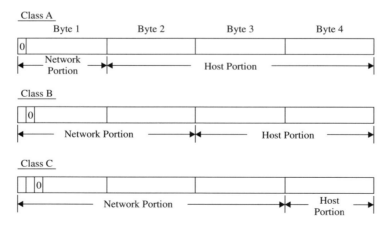

Figure 3.4   Classful IPv4 address formats.

address of 127 is the well-known loopback address, reducing available Class A addresses to a maximum value of 126.

We can note the general structure of a Class A address as follows, where N represents a network byte and H represents a host file.

$$\langle N \rangle \langle H \rangle \langle H \rangle \langle H \rangle$$

When focusing our attention upon Class B and Class C addresses we should note that each class increases the number of bytes used for the network while decreasing the number of bytes used to define the host on the network. Thus, the general structure of a Class B address is as follows:

$$\langle N \rangle \langle N \rangle \langle H \rangle \langle H \rangle$$

Similarly, we can denote the general structure of a Class C address as follows:

$$\langle N \rangle \langle N \rangle \langle N \rangle \langle H \rangle$$

A Class B address is identified by the setting of the first two bits in the first byte to a binary value of 10. In comparison, a Class C address is identified by the setting of the first three bits in the first byte to a binary value of 110.

Due to the growth in the use of the Internet, Class A, B and C addresses, commonly referred to as classful addresses, became a rare resource. Since many organizations need to connect two or more networks to the Internet the assignment of a classful address for each network could result in a waste of address space. For example, consider the Class C network address of 198.78.46.0. Here this Class C address permits 256 unique hosts (0 through 255) to be identified. However, because a host address of 0 could be confused with the basic network address and 255 (hex FF or all 1's) is the broadcast address, the maximum number of distinct hosts on a Class C network is reduced by 2 to 254. If an organization had two networks, each with 20 hosts, the assignment of two Class C network addresses would waste $254 * 2 - 2 * 20$, or 468 addresses. Recognizing that a method to conserve address space was desirable resulted in the development of subnetting and the use of the subnet mask.

3.2.2 Subnetting

Subnetting can be viewed as the process of subdividing a classful address into separate entities. From the Internet routers are not aware of the subdivision; however, upon receipt of a datagram the subnet mask configured on both routers and on hosts are used to correctly route datagrams onto an applicable subnet as well as enable hosts on the subnet to recognize datagrams addressed to them.

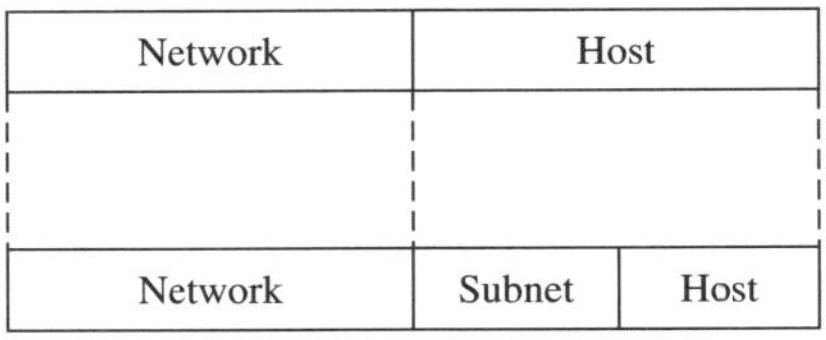

Network	Host	
Network	Subnet	Host

Figure 3.5 The subnetting process results in the conversion of a two-level classful address into a three-level address.

You can view subnetting as the process of converting a two-level (network and host) address into a three-level address consisting of network, subnet and host address portions. Figure 3.5 illustrates the subnetting process.

In examining the relationship between the two-level and three-level addresses shown in Figure 3.5 note that the subnet identifier is taken from the host portion of the classful IP address. This explains why subnetting is not recognized on the Internet since routing occurs based upon the network portion of the address. Also note that as you increase the number of bit positions allocated to identify a subnet, you decrease the number of bit positions allocated to the host positions of the address.

As an example of the use of subnetting, assume you have five networks located in a building, each with 20 hosts, and your organization was issued a single Class C address. Because two bits can only identify four subnets, you would need to use three bit positions for the subnet portion of the address, enabling 23, or eight unique subnets to be identified. Assuming the network used is 198.78.46.0, then the relationship between the network, subnet and host portions of the Class C address could be indicated as shown in Figure 3.6.

```
Network     198  .  78  .  46  .   0

Subnet 0    11000000.01001110.00100110.000xxxxx

Subnet 1    11000000.01001110.00100110.001xxxxx

Subnet 2    11000000.01001110.00100110.010xxxxx

Subnet 3    11000000.01001110.00100110.011xxxxx

Subnet 4    11000000.01001110.00100110.100xxxxx

Subnet 5    11000000.01001110.00100110.101xxxxx
```

Figure 3.6 Relationship between the network, subnet and host portions of the Class C address 198.78.46.0.

For each subnet the host is represented by five bits, resulting in 32 distinct values that range from 00000 to 11111. However, similar to classful network addresses, you cannot have a host address of all zeros or all 1's on a subnet. Thus, for each five-bit subnet the acceptable host numbers range from 00001 (decimal 1) to 11110 (decimal 30) for a total of 30 possible distinct hosts.

Based upon the preceding, the host addresses on subnet 0 will range from 00001 to 11110, while the fourth dotted decimal number in the address will range from 1 to 30. Continuing our adventure in subnetting, the subnet prefix for subnet 1 is 001. Since hosts on that subnet will have addresses from 00001 to 11110, then the full 8-bit byte range becomes 00100001 to 00111110. Thus, on subnet 1 the host's dotted decimal range becomes 33 through 62.

3.2.3 The subnet mask

Both routers and hosts are configured with subnet masks as a mechanism to route and recognize datagrams destined to a particular subnet. The subnet mask is formed by using a sequence of binary 1's to extend the network portion of a classful address through the subnet series of bits. Fore example, using the three-bit position subnet shown in Figure 3.6 would result in a subnet mask as follows:

$$11111111.11111111.11111111.11100000$$

The subnet mask in dotted decimal notation would then be entered as 255.255.255.224. Upon receipt of a datagram containing the network address 198.78.46.0 a router would note that the first two bits in the network portion of the address are set, indicating that the address is a Class C address. This indicates that the network portion of the address is contained in the first 24 bits, while the host address is contained in the last 8 bits of the 32-bit address. Using the subnet mask, the router notes it is 27 bits in length. Because the router knows it is working with a Class C address for which the network portion of the address is 24 bits, this informs the router that the first three bits in the fourth byte indicate the position of the subnet mask. Thus, the router can transfer the datagram onto the correct subnet. Now that we have an appreciation for the subnet mask, let's turn our attention to the wildcard mask.

3.2.4 The wildcard mask

The wildcard mask represents an identifier used in Cisco access lists to define matches and don't care conditions. In a Cisco access list you can specify an IP address to be permitted or denied the ability to flow through an interface via the use of a permit or deny statement. The format of a standard Cisco IP access list is shown below:

access-list[list number][permit|deny][source address][wildcard mask]

Here the list number would be between 1 and 99 to identify a standard IP access list. All statements in an access list would be configured with the same list numbers. The keyword 'permit' or 'deny' would define whether datagrams with the defined source address are allowed to flow through the router interface or are passed to the big bit bucket in the sky. For a standard IP access list the source address can represent a specific IP address of a host or a group of hosts. To specify a group of hosts requires the use of the wildcard mask, which brings us back to the topic of this section.

The wildcard mask is used in conjunction with an IP address to define a host or group of hosts. To do so the mask uses a binary 0 to represent a match, while a binary 1 is used to represent a don't care condition. Returning to our use of the 198.78.46.0 network, if you wanted to allow all hosts on that network, the required wildcard mask would be 0.0.0.255. Here the first three 0's indicate we want to match the network portion of the address, while the value of 255 decimal in the fourth byte is equivalent to 11111111 binary, which indicates we don't care about the host address. Thus, the standard IP access list statement required to allow all packets with a source address of 198.78.46.0 to flow through a router's interface would be:

access-list 1 permit 198.78.46.0 0.0.0.255

Note that the wildcard mask is the complement of the subnet mask. Thus, another technique you can use to specify an access list wildcard mask is to first determine the subnet mask and then take the inverse.

Now, let's consider the network diagram illustrated in Figure 3.7 and assume the access list is applied to the serial 0 (S0) interface of the router located in the top left portion of the referenced figure in the inbound direction. This simple 1 line access list can be considered as establishing an elementary VPN since it only allows traffic from a predefined network to flow through the router. At the opposite end of the Internet connection you would establish a similar access list to enable datagrams from the 205.131.175.0 network to flow through the router. Doing so in effect establishes an elementary VPN in the opposite direction. Although each of the router access list statements by themselves do not result in the encryption of the contents of datagrams, they control the flow of packets through each router.

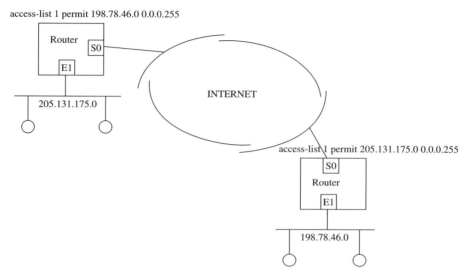

access-list 1 permit 198.78.46.0 0.0.0.255

access-list 1 permit 205.131.175.0 0.0.0.255

Figure 3.7 Establishing an elementary VPN through the use of standard IP access lists.

3.2.5 ICMP

As noted earlier in this chapter, ICMP messages are transported as IP datagrams with an IP header prefixed to each message. Each ICMP message begins with three common fields, with the remaining fields in a particular message structured based upon the specific message. The three common fields include an 8-bit Type field that defines the ICMP message, an 8-bit Code field that may provide additional information about a particular message type, and a 16-bit Checksum field that provides integrity for the message.

Table 3.3 lists 13 defined and actively used ICMP message types and their type field values. In this section we will briefly examine the function and use of each ICMP message. In addition, we will also note the applicability of ICMP messages to a VPN networking environment.

3.2.5.1 Echo Request and Echo Reply

The ICMP echo request (type 8) and echo reply (type 0) messages are used to test whether a destination is active and reachable. A host or router will transmit an echo request to a distant device. That device, if both active and reachable, will respond with an echo reply. Both echo request and echo reply ICMP messages are used by the well-known Ping application.

TABLE 3.3 ICMP type field

Type field value	Defined ICMP message type
0	Echo Reply
3	Destination Unreachable
4	Source Quench
5	Redirect
8	Echo Request
11	Time Exceeded
12	Parameter Problem
13	Timestamp Request
14	Timestamp Reply
15	Information Request
16	Information Reply
17	Address Mask Request
18	Address Mask Reply

Many organizations by default block all or most ICMP messages flowing through their router or firewall. If this situation occurs, you obviously cannot use the Ping application to determine if a destination is active and reachable. Thus, any testing associated with a VPN may require the reconfiguration of your routers and firewalls to enable echo request and echo reply ICMP messages to flow through those devices. Both echo request and echo response messages do not have a qualifier, resulting in their Code field value being set to zero.

3.2.5.2 Destination Unreachable

One of the more important ICMP messages for determining the reason why an IP datagram cannot reach its destination is a type 3, destination unreachable message. The Code field in a Type 3 message further qualifies the reason why the message could not be delivered to its intended recipient. Table 3.4

TABLE 3.4 Destination unreachable code field values

Code field value	Meaning
0	Network Unreachable
1	Host Unreachable
2	Protocol Unreachable
3	Port Unreachable
4	Fragmentation Needed, OF Bit Set
5	Source Route Failed

lists the Code field values and their meanings for a destination unreachable message.

In examining the entries in Table 3.4 it is important to note that routers will transmit a network or host unreachable message when they cannot route or deliver an IP datagram. The code field value clarifies the reason why the datagram could not be delivered and can be valuable for determining or isolating network-related problems. If your organization by default prohibits ICMP messages to flow through your security devices, you may need to reconfigure those devices if datagrams cannot reach their destination. By allowing ICMP Type 3 destination unreachable messages to flow through security devices, you can obtain information concerning why datagrams could not be delivered.

3.2.5.3 Source Quench

An ICMP source quench message represents a method used by routers and hosts to control the flow of data. That is, when datagrams arrive at a rate higher than the processing rate of a router or host, they will discard them. The device that discards the datagrams will transmit an ICMP source quench message, in effect informing the original source to slow down its rate of sending datagrams. Typically, routers and hosts will transmit one source quench message for each datagram they discard. Similar to echo request and echo response messages, there are no code field qualifiers for the source quench message. This results in the message having its Code field value set to zero. Although rarely used today, in a VPN environment it is quite common for the encryption and decryption functions performed by certain hardware to delay the flow of datagrams. When the buffer of the device fills up, additional datagrams arriving are discarded. In this situation the flow of source quench message might be warranted.

3.2.5.4 Redirect

When a router detects the fact that a host is using a non-optimum route it will send an ICMP redirect message to the host. While the intentions for the use of this message are beyond reproach, because it can be used by hackers, it is commonly blocked by routers and firewalls as a matter of organizational policy. In a VPN environment it is highly doubtful if this ICMP message is required.

3.2.5.5 Time Exceeded

The ICMP time exceeded message is generated by a router when it has to discard a datagram. As we noted earlier in this chapter when we discussed the IP header, the Time-to-Live (TTL) field value is decremented each time a datagram flows through a router. When the TTL value reaches zero, a router both discards the datagram and sends an ICMP time exceeded message back to the source address in the datagram. A second reason for a datagram to be discarded occurs when fragment reassembly time is exceeded. When fragmentation occurs, a receiving host will initiate a timer upon receipt of the first fragment transported in a datagram. If the timer expires prior to the arrival of all fragments, an ICMP time exceeded message is returned to the originator. To differentiate the two timeouts the Code field value in the ICMP message is used. That is, a code field value of 0 indicates a time-to-live count value was exceeded while a code field value of 1 indicates the fragment reassembly time was exceeded.

In a VPN environment where IP datagrams are tunneled, the original TTL field is not affected as packets flow across a router boundary. Thus, you cannot use the inner heading as a mechanism to track the flow of packets through routing devices. Instead, you need to track the outer header. In Chapter 5 when we turn our attention to IPSec, we will note the relationship of inner and outer headers as packets are tunneled.

3.2.5.6 Parameter Problem

When a router or host encounters problem interpreting the fields with an IP header, it will return an ICMP parameter problem message to the source. This message includes a pointer that identifies the byte in the header that caused the problem. Normally, this ICMP message is not required in a VPN environment.

3.2.5.7 Timestamp Request and Reply

ICMP message types 12 and 13 represent Timestamp Request and Timestamp Reply messages. These two messages are used to synchronize the clocks of two devices. In addition, the fields within the two messages can be used to estimate the transit time of datagrams between two devices. For most VPN operations it is doubtful if these ICMP messages will be required.

3.2.5.8 Information Request and Reply

Another pair of ICMP messages are message types 15 and 16, better known as Information Request and Information Reply. The ICMP Information Request message is used by devices to obtain an IP address for a network to which they are attached and functions as an alternative to the use of a reverse ARP. The Information Reply transports the response to the Information Request. Similar to the ICMP timestamp request and reply, there is typically no need for ICMP information request and information reply messages in a VPN environment.

3.2.5.9 Address Mask Request and Reply

Another pair of ICMP messages are types 17 and 18, address mask request and address mask reply. The purpose of this pair of ICMP messages is to allow a device to learn its subnet mask. To do so, a device will transmit an ICMP address mask request message to a router. If the device was not previously configured with the IP address of the router, it can broadcast the message. For either situation the router will return the address mask in an ICMP address mask reply message. Similar to the two previously discussed ICMP message pairs, in a modern VPN environment it is doubtful if you would need to allow ICMP type 17 and type 18 messages to flow through routers and firewalls. Now that we have an appreciation for ICMP as it relates to VPNs, let's move up the protocol stack and turn our attention to the transport layer.

3.3 The transport layer

In the TCP/IP protocol suite the transport layer permits multiple applications to flow to a common destination, either from the same source or from different data originators. The transport layer resides between the network layer and the application layer, receiving application data, encapsulating the data with a transport header that identifies the type of application, and providing the encapsulated data and header to the network layer for transmission onto the network.

3.3.1 Transport layer protocols

Included in the TCP/IP protocol suite are two commonly used transport layer protocols – the Transmission Control Protocol (TCP) and the User Datagram Protocol (UDP). TCP represents a connection-oriented, reliable transport protocol that creates a virtual circuit for the transfer of information. In comparison, UDP represents a connectionless, best-effort transport layer protocol.

3.3.2 The TCP header

The top portion of Figure 3.8 illustrates the TCP header. Included in the TCP header are fields that are used to ensure datagrams are received correctly with respect to both their content and sequence. Another important function of the TCP header is to denote the type of application data carried by each

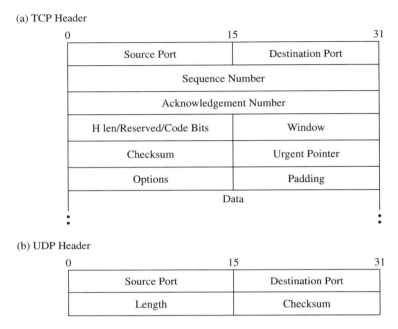

(a) TCP Header

(b) UDP Header

Figure 3.8   The TCP and UDP headers.

datagram. This function is accomplished by the use of the port fields that identify the process or application transported in the datagram. In actuality, the TCP header plus application data is referred to as a segment, resulting in the port number identifying the type of data in the segment. When the IP header is prefixed to the TCP segment, the resulting datagram contains the source and destination IP addresses and enables the segments to be delivered via the network. For the purposes of this book, which is focused upon VPNs, we will primarily limit our discussion of the TCP and UDP headers to their source and destination port fields.

3.3.3 The UDP header

The UDP header is shown in the lower portion of Figure 3.8. In comparing the UDP header to the TCP header it is obvious that the former is streamlined with respect to the latter. This streamlining results from the fact that UDP is a best-effort protocol that depends upon higher layers for error detection and correction, which enables most of the TCP fields to be dropped from use. However, similar to TCP, the UDP header includes 16-bit source and destination ports that identify the process or application. Thus, let's look at those common ports.

3.3.4 Source and destination port fields

The source and destination port fields are each 16 bits in length. The source port field contains a port number that theoretically denotes the application associated with the data generated by the originating station. The reason the term 'theoretically' was used is that in most transmissions the source port number is randomly generated by the originator. If the source port is not used, its value is set to 0. In comparison, the destination port field contains a port number that identifies a user process or application for the receiving station, enabling it to distinguish different applications transmitted from a common location. For example, when a station initiates a file transfer, it might open FTP to transfer data using port number 1234 as the source port while later in the day a second file transfer might occur with the station using source port 2345. However, for all FTP transfers the destination port would be fixed at 21, which is the standard port number for which FTP incoming data is received. Because FTP is transported using TCP, when the destination station receives the incoming data, it responds by creating a segment and placing the source port number in the destination port field. This action enables the file originator to correctly identify the response to the datagram. Because there are three types of port numbers that can be used in the port fields, and both TCP and UDP headers have the same 16-bit source and destination port fields, let's turn our attention to this topic.

3.3.4.1 Port Numbers

Both TCP and UDP headers contain 16-bit source and destination port fields, permitting port numbers in the range 0 through 65535. This results in 65536 distinct port numbers being available for use. This 'universe' of port numbers is subdivided into three ranges, referred to as well-known ports, registered ports, and dynamic or private ports.

3.3.4.2 Well-Known Ports

Well-known ports are also referred to as assigned ports, as their assignment is controlled by the Internet Assigned Numbers Authority (IANA). Well-known or assigned ports are in the range 0 through 1023 and are used to indicate the transportation of standardized processes. Where possible, the same well-known port number assignments are used with TCP and UDP. Ports used with

TCP commonly provide connections that transport relatively long-term connections, such as file transfers and remote access. In the early literature references to well-known port numbers are specified as being in the range of values from 0 through 255. While that range was correct many years ago, the range for assigned ports managed and controlled by the IANA is now from 0 to 1023.

3.3.4.3 Registered Ports

Assigned port numbers range from 0 to 1025 out of the universe of 65536 available ports. Port numbers that exceed 1023 can be used by any process or application; however, doing so in a haphazard manner could create incompatibilities between vendor products. Recognizing this potential problem, the IANA permits vendors to register the use of port numbers. The result is the use of the range of port number values from 1024 through 49151 for registered ports. Although an application or process may be registered, its registration does not hold legal implications, and it is primarily to enable other vendors to develop compatible products as well as to enable end-users to set up equipment appropriately. For example, if a new application uses a registered port number, it is relatively easy to adjust a router access list or firewall configuration to enable the flow of datagrams containing the new application. Although developers can use any port number beyond 1023, many respect registered port numbers.

3.3.4.4 Dynamic Ports

Dynamic port numbers are in the range 49152 through 65535. Port numbers in this range are typically used by vendors implementing proprietary network applications, such as a method to transport digitized voice for a specific application unique to a vendor. Another common use of dynamic port numbers includes the random selection of a port number by certain applications.

Table 3.5 provides a summary of some of the more popular well-known and registered port numbers. In examining the entries in Table 3.5 note that Secure HTTP, which in effect provides a client–server VPN for Web browsing, is defined by a well-known port. Similarly, Kerberos and Socks are also defined through the use of a well-known port. In comparison, the point-to-point tunneling protocol is defined through the use of a registered port. Although many protocols used to transport VPN traffic are defined through port usage, IPSec provides authentication and encryption that results in new headers identified by the protocol field in the IP header. Thus, as we probe further into VPNs we will note they can be identified in several ways.

TABLE 3.5   Examples of well-known and registered TCP and UDP services and port utilization

Service	Port type	Port number
(Well-known ports)		
Remote Job Entry	TCP	5
Echo	TCP and UDP	7
Quote of the Day	TCP	17
File Transfer (Data)	TCP	20
File transfer (Control)	TCP	21
Secure Shell	TCP	22
Telnet	TCP	23
Simple Mail Transfer Protocol	TCP	25
Domain Name Server	TCP and UDP	53
Trivial File Transfer Protocol	UDP	69
Finger	TCP	79
Hypertext Transfer Protocol	TCP	80
Kerberos	TCP	88
Lightweight Directory Access Protocol	TCP and UDP	389
Secure HTTP	TCP	443
Socks	TCP	1080
(Registered ports)		
Point-to-Point Tunneling Protocol	TCP	1723
MSN Messenger	TCP	1863
Yahoo Messenger-Voice Chat	TCP and UDP	5000-5001
Yahoo Messenger-Messages	TCP	5050
Yahoo Messenger-Webcams	TCP	5100
AOL Instant Messenger	TCP	5190

3.4 Proxy services and network address translation

In concluding this chapter we will turn our attention to two interrelated topics that are relevant to VPN operations: proxy services and network address translation (NAT).

3.4.1 Proxy service

A proxy service represents a service that works on behalf of another party. For example, a WWW proxy would provide caching of Web pages on behalf of network users. When a client requests a page previously viewed by another network user, the proxy server would provide the new requester with the previously cached page, enhancing performance while minimizing traffic over the wide area network. One common proxy service is network address translation.

3.4.2 Network address translation

Network address translation can be performed by several types of hardware devices to include routers and firewalls. NAT was originally developed as a mechanism to extend the use of scarce IPv4 address space. As the use of the Internet expanded, the ability of organizations to obtain registered IP addresses from their service providers became more difficult. Recognizing the fact that only a small portion of local network users would be accessing the Internet at a particular point in time, it became possible for organizations to assign each station a private IP address, typically from one of the three blocks of addresses reserved in RFC 1918. Those address blocks are listed in Table 3.6.

By using an address translator to map or translate unregistered private IP addresses into registered addresses on the other side of a device, it became possible to have more devices than registered addresses. For example, if an organization has 1000 stations, the mapping of 1000 private unregistered IP addresses to the 254 addresses in a Class C network enables one Class C network address to be used instead of four such addresses. However, if more than 254 users simultaneously required Internet access, some user requests must be either denied or queued until a previously used registered address becomes available. Although NAT was primarily developed as a technique to conserve difficult-to-obtain IPv4 addresses, a side benefit of its use is the fact that it hides the addresses of stations behind the translator. This means that a direct attack upon the organizational hosts is no longer possible, and resulted in NAT functionality being incorporated into firewalls in addition to its use in routers.

Regardless of the device used to perform NAT, its operation is similar. That is, as datagrams arrive at the device performing NAT, the source address is translated into a public address for transmission onto the Internet. In

TABLE 3.6 Reserved IPv4 addresses for private Internet use (RFC 1918)

Address Blocks
10.0. 0.0—10.255.255.255
172.16. 0.0—172.31.255.255
192.168.0.0—192.168.255.255

comparison, inbound datagrams have their public IP address translated into their equivalent private IP address based upon the state of an IP address mapping table maintained by the device.

3.4.3 Types of address translation

There are three types of NAT that devices can employ. These types or methods of address translation are static NAT, pooled NAT, and port-level NAT, with the latter also referred to as Port Address Translation (PAT).

3.4.3.1 Static NAT

Static NAT results in the permanent mapping of each host on an internal network to an address on an external network. Although static mapping does not provide a reduction in the number of IP addresses needed by an organization, after it has been configured, no further action is necessary and it occurs via a simple table lookup process which minimizes processing delay.

3.4.3.2 Pooled NAT

When a pooled NAT technique is used, a pool of addresses on the external network is used for the dynamic assignment of IP addresses in place of the use of private addresses on the internal network. Although pooled NAT enables users to conserve the use of public IP addresses, its use can adversely affect certain types of applications. For example, SNMP managers track devices based upon the device IP address and an object identifier. Because pooled NAT means that network addresses will more than likely change over time, this means that devices in front of the translating device cannot be configured to reliably transmit SNMP traps to devices behind the translating device. One possible solution to this problem is to permanently map an SNMP manager to an IP address while all other devices share the remaining addresses in the address pool. Of course, the device that supports pooled NAT must also be capable of permitting support for static mapping.

3.4.3.3 Port Address Translation

A third type of address translation results in the mapping of internal addresses to a single IP address on the external network. To accomplish this, the address translator assignees different port numbers to TCP and UDP source ports. The port numbers used for mappings are those above 1023, providing 64512 (65535 − 1023) simultaneous TCP/IP or UDP/IP connections on a single IP address. Because mappings occur on a single IP address through the use of different port numbers, this technique is referred to as Port Address Translation (PAT). The use of PAT results in all traffic transmitted onto the public network appearing to come from a single IP address.

3.4.4 VPN considerations

The use of NAT with a VPN can result in some interesting problems. One such problem occurs when you use a VPN protocol that periodically generates keep-alive frames during periods of no activity. If a pooled address translation is used and a timeout value results in a new address assigned to one side of the VPN, the other side issuing keep-alives will be none the wiser. This action would result in keep-alives failing to flow to the other side of the VPN. After a period of time with no activity to include keep-alives, the VPN connection would terminate. The solution to this problem is either to use NAT translation or employ static mapping for VPN devices.

A second area of concern with respect to NAT and VPNs occurs when VPN gateways are behind a network address translation device. When this situation occurs, a number of VPN connections can be adversely affected if a pooled address is associated with a gateway and the address changes. Again, the solution to this problem is the use of static IP addresses or the use of NAT.

A third area of concern with respect to NAT and VPNs occurs if an address translation device resides between clients and a VPN gateway performing a method of IPSec referred to as tunnel mode operations. Under tunnel mode operations the VPN gateway prefixes a new IP header to the original IP header generated by a client, enabling authentication and encryption of data. Unfortunately, if the NAT device resides between the clients and the VPN gateway, once again it is possible for a change in the assignment of the IP address in a pooled NAT environment to result in unexpected consequences. For example, assume the NAT device changes the addresses of the client. The gateway is none the wiser and would continue to receive datagrams from the opposite end of the connection with imbedded headers directed to the previously used address.

A fourth area of concern with respect to NAT and VPNs concerns the use of the tunnel mode. When using the tunnel mode, the inner IP datagram remains as is, with the outer IP header operated upon by NAT. This means you cannot interconnect two networks that use the same RFC 1918 private network addresses as doing so would cause routing problems on each network. Instead, you would have to change the network address on one of the interconnected networks.

When working with VPN gateways it is a good idea to attempt to avoid a pooled NAT environment. In addition, when your organization requires both VPN gateways and a NAT capability, you should consider acquiring products that perform both operations as well as support configuration flexibility, as doing so can be expected to alleviate the previously described problems.

chapter four

Layer 2 Operations

The aim of this chapter is to obtain an understanding of layer 2 VPN operations. In this chapter we will focus our attention upon the Point-to-Point Tunneling Protocol (PPTP), Layer 2 Forwarding (L2F), and the Layer 2 Tunneling Protocol (L2TP). Because we will discuss IPSec in Chapter 5, we will defer a discussion of the use of IPSec with L2TP until then. Because the tunneling protocols that are covered in this chapter are based upon the Point-to-Point Protocol (PPP), we will review that protocol first.

4.1 The Point-to-Point Protocol

The Point-to-Point Protocol (PPP) was developed to provide a mechanism to transport multi-protocol datagrams over point-to-point serial links. PPP supports both asynchronous and synchronous operations and is defined in RFC 1661, which made RFC 1548 obsolete. If you are familiar with HDLC, you will note that the structure of PPTP is very similar to that protocol. This is by design as PPTP was derived from HDLC.

4.1.1 Components

There are three primary components of PPP. These components are a mechanism that encapsulates multiprotocol datagrams, a link control protocol (LCP) and a family of network control protocols (NCPs). The LCP is used to establish, configure and test the data link connection. For example, upon the initialization of a connection, the LCP will define the method of encapsulation to be used as well as define agreements between the two parties concerning how to handle varying packet lengths and different types of error conditions. Included with the LCP are optional facilities to authenticate the identity of

Virtual Private Networking   G. H. Held
© 2004 John Wiley & Sons, Ltd   ISBN: 0-470-85432-4

the peer on the connection as well as determine when a link is working correctly and when it is failing, with information concerning the latter two events based upon periodic testing of the connection. Once a PPP session is completed, the LCP becomes responsible for terminating the connection in an orderly manner.

In comparison to one LCP, a family of NCPs is defined under the PPP standard. Each NCP is designed to provide a negotiation and management function for a specific network layer protocol being transported and is defined in companion documents to RFC 1661. For example, NCPs are defined for such layer 3 protocols as IP, IPX, DECnet and AppleTalk.

4.1.2 PPP encapsulation

Figure 4.1 illustrates the format of a PPP encapsulated datagram. Unlike an IP datagram which is prefixed with an IP header followed by a transport layer header, the format shown in the upper section of Figure 4.1 is conspicuous by the absence of those headers. In actuality, those headers exist in a TCP/IP environment when PPP is transporting IP datagrams, with the datagrams transported in the information field as indicated in the lower portion of the illustration. Because PPP is a layer 2 transport facility, it only transports data between pairs of stations and, as we will shortly note, uses the Protocol field to define the type of data transported.

In examining Figure 4.1 you will note the PPP frame consists of six fields. The first three fields, Flag, Address and Control, represent framing information, while the Protocol field identifies the type of information transported in the variable Information field. The sixth field, FCS, provides an error detection and correction capability. Now we will probe deeper into each field.

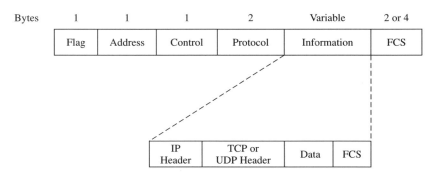

Figure 4.1 The format of a PPP packet transporting an IP datagram.

4.1.2.1 Flag Field

The Flag field is 8 bits in length and serves to indicate the beginning and end of a packet. The Flag field has the value 01111110, which is the same value used in the HDLC flag field. To protect data from being misinterpreted as a flag, any sequence of five set bits is modified so that a binary 0 is inserted. Similarly, any received sequence of five binary 1's will have the previously inserted binary 0 removed, a process referred to as zero insertion and removal. Because the composition of the flag is then unique, it will indicate the beginning of a packet, which in effect is the end of the preceding packet.

4.1.2.2 Address Field

The Address field is a single byte whose value is always set to hex FF or all 1's, which is the broadcast address. Because PPP transmits data between two devices, there is no need to assign individual station addresses. This explains why a broadcast address is used instead of a specific address.

4.1.2.3 Control Field

This one-byte field contains the value hex 03. This value specifies transmission of user data in unsequenced frames. Because PPP occurs over a serial link connecting two locations, the protocol does not support unsequenced frames and does not include a sequence number field.

4.1.2.4 Protocol Field

The two-byte Protocol field identifies the protocol that is encapsulated within the Information field of the packet. Unlike the IP header Protocol field where field values appear to be assigned randomly, the equivalent PPP field has structured values. For example, field values in the hex 0**X to 3*** range are used to identify the network layer protocol of specific packets. In comparison, Protocol field values in the hex 4*** to 7*** range are used for protocols with low volume traffic that have no associated NCP. Protocol field values in the range hex 8*** to C*** identify packets belonging to the associated network control protocol, while Protocol field values in the range hex C*** to f*** identify packets as link-layer control protocols (LCPs). Table 4.1 provides a summary of PPP Protocol field values and their meanings.

In examining Protocol field values contained in Table 4.1, note that they are all odd. This is by design in accordance with ISO requirements. A full list of Protocol field values are defined in RFC 1700, titled 'Assigned

TABLE 4.1 PPP Protocol Field Values

Value in Hex	Meaning
0001	Padding Protocol
0003-001f	Reserved (Transparency Inefficiency)
0021	Internet Protocol
002a	AppleTalk
002b	Novell IPX
002d	Van Jacobson Compressed TCP/IP
002f	Van Jacobson Uncompressed TCP/IP
0035	Banyan VINES
003f	NetBIOS Framing
0041	Cisco Systems
004f	IPv6 Header Compression
8021	Internet Protocol Control Protocol
802a	AppleTalk Control Protocol
802b	Novell IPX Control Protocol
8031	Bridging NCP
8035	Banyan VINES Control Protocol
803f	NetBIOS Framing Control Protocol
8041	Cisco Systems Control Protocol
804f	IPv6 Header Compression Control Protocol
C021	Link Control Protocol
C023	Password Authentication Protocol
C026	Link Quality Report
C223	Challenge-Handshake Authentication Protocol

Numbers.' Based upon our prior discussion of Protocol field code ranges, we can see that from our list in Table 4.1 the protocols defined in the lower code range of hex 0000–02ff that identify network layer protocols have corresponding link layer control protocols in the range hex C000–ffff. For example, the Protocol field code value of 0021 defines the Internet Protocol, while a code field value of 8021 defines the Internet Protocol Control Protocol. Also note that two authentication methods are defined as link layer control protocols, code C023 for the Password Authentication Protocol (PAP) and code C223 for the Challenge-Handshake Authentication Protocol (CHAP).

As a quick review, the key advantage of CHAP over PAP is the fact that CHAP in effect transmits an encrypted password that cannot be detected by a third party operating a network analyzer. In comparison, the PAP password is transmitted in the clear and can easily be noted by an unauthorized third party.

4.1.2.5 Information Field

This variable length field is used to transport a datagram for the protocol specified in the Protocol field. The default maximum length of this field is 1500 bytes; however, by prior agreement during the network layer configuration negotiation process, a different maximum information field length can be selected.

Because this field is variable in length and not prefixed with a length field, a trick is used to determine the end of the field. The receiver first looks for the next flag field, which indicates the end of the packet and beginning of the next packet. By allowing two bytes for the Frame Check Sequence (FCS) field the receiver is able to note the end of the previously received Information field.

4.1.2.6 Frame Check Sequence Field

The Frame Check Sequence (FCS) field provides a mechanism for a receiver to detect the occurrence of one or more bit errors in the received packet. Normally the FCS field is 2 bytes in length; however, during link negotiation a 4-byte FCS can be selected for improved error detection.

4.1.3 Link control protocol operations

Prior to actual user information being able to flow across a PPP link, the two endpoints will test the link as well as negotiate configuration parameters. Once this is accomplished, network layer protocol configuration negotiation can occur, resulting in the exchange of user information. If LCP has closed the link, it will inform the network layer protocols, enabling them to take appropriate action. Assuming all data is exchanged, LCP will terminate the link at the end of the exchange; however, it can also terminate the link at any time. Because LCP is the key for establishing, controlling, maintaining and terminating a point-to-point connection, let's examine how it operates.

4.1.3.1 LCP Packet Format

Figure 4.2 illustrates the Link Control Protocol packet format. The first four fields, Flag, Address, Control and Protocol, have previously been described. The value of the Protocol field would be set to hex C021 to indicate that the packet is a Link Control Packet. The remainder of the packet consists of five fields, of which the FCS field was previously described. Thus, we will

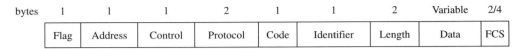

bytes	1	1	1	2	1	1	2	Variable	2/4
	Flag	Address	Control	Protocol	Code	Identifier	Length	Data	FCS

Figure 4.2   The Link Control Protocol packet format.

first focus our attention upon the Code, Identifier, and Length fields prior to discussing how LCP packets operate.

Code field

The Code field, which is 1 byte in length, identifies the type of link control packet. Table 4.2 lists Code field values in hex and their meanings.

Identifier field

The Identifier field is 1 byte in length. This field is used to match requests and replies. When a packet is received with an invalid Identifier field value, it is simply discarded, which in the literature is referred to as being silently discarded.

Length field

The Length field is 2 bytes in length. The value of this field specifies the length of the LCP packet to include the Code, Identifier, Length and Data fields. The value of the Length field cannot exceed the Maximum Receive Unit (MRU) of the link.

TABLE 4.2   Link Control Packet Code types

Code (Hex)	Meaning
01	Configure-Request
02	Configure-Ack
03	Configure-Nak
04	Configure-Reject
05	Terminate-Request
06	Terminate-Ack
07	Code-Reject
08	Protocol-Reject
09	Echo-Request
0a	Echo-Reply
0b	Discard-Request

4.1.3.2 Packet Classes

By examining the Code field values listed in Table 4.2 it is relatively easy to see that LCP packets can be subdivided into three distinct classes. One class involves the use of LCP packets to establish and configure a link. Packets in this class include a Configure-Request packet that is used to open a connection and a Response packet that is sent in response to the request. Here the response will be either a Configure-Ack, Configure-Nak, or a Configure-Reject packet.

A second class of LCP packets involves those used to manage and test a connection. Packets within the link maintenance class of packets include Code-Reject, Protocol-Reject, Echo-Request, Echo-Reply and Discard-Request packets.

The third class of LCP packets involves those associated with the termination of an existing connection. There are two types of packets that fall into the link termination class, Terminate-Request and Terminate-Ack. Now that we have an appreciation for the general relationship of currently defined LCP packets, let's probe deeper and observe how they interact.

4.1.3.3 Connection Establishment

As previously mentioned, one class of LCP packets are used to establish a connection. To do so LCP will transmit a Configure-Request packet. Because there are two endpoints in a PPP system and transmission is bi-directional, each endpoint in effect transmits a Configure-Request packet. That packet contains the desired values of the endpoint for PPP operating parameters. If the opposite endpoint agrees, it will return a Configure-Ack. If the opposite endpoint does not agree to the initial configuration values for PPP operations, it will return a Configure-Nak. The Configure-Nak will convey the unacceptable configuration options to the originator. The originator can then respond with alternative configuration values. If, in the event one or more configuration options received in a Configure-Request are not recognizable or are not acceptable for negotiation due to the configuration of the receiving endpoint, that endpoint will respond with a Configure-Reject. Thus, the response to a Configure-Request packet can be either a Configure-Ack, Configure-Nak or a Configure-Reject.

Figure 4.3 illustrates a PPP configuration negotiation between two endpoints. In examining Figure 4.3 we will assume that the first Configure-Request packet contained one or more parameters that endpoint 2 could not support. This resulted in endpoint 2 responding with a Configure-Nak that contains alternative values for one or more parameters contained in the initial

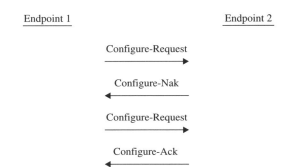

Figure 4.3   An example of the PPP link establishment configuration negotiation process.

Configure-Request packet. Assuming the first endpoint adopts those values in its next Configure-Request LCP packet, endpoint 2 now responds with a Configure-Ack.

For simplicity of illustration Figure 4.3 shows only half of the negotiation process, indicating the sequence of negotiation from the view of endpoint 1. A parallel negotiation process will occur from endpoint 2 to endpoint 1. Because the time in the negotiation process precludes the transfer of user data a series of default values were established for all configuration options. An endpoint willing to accept default values for one or more options does not have to propose a value for such options. Thus, if both endpoints agree to all the defaults in their initial Configure-Request packets, the negotiation will be rapidly completed with the responding Configure-Ack packet.

4.1.3.4 Link Maintenance

As noted earlier in this chapter, we can classify five LCP packets as being suitable for link maintenance operations. The Code-Reject packet is issued by an endpoint in response to the receipt of an LCP packet with an unknown value in its Code field. The invalid Code field value can result either from a data error or from the endpoint attempting to use a code not supported by the other endpoint. In comparison, a Protocol-Reject packet is issued when an endpoint receives a PPP packet with an unknown Protocol field. Here the receipt of a packet with an unknown Protocol field value could either represent an error or the fact that the other endpoint was attempting to use a protocol that the receiving endpoint does not support. For either situation the receiving endpoint responds with a Protocol-Reject packet.

There are two link control protocol maintenance packets that are used in pairs. An Echo-Request packet transmitted by one endpoint will result in the other endpoint responding with an Echo-Reply packet. The fifth type of link maintenance packet is the Discard-Request. This LCP packet is transmitted as a mechanism to exercise the local to remote direction of the communications link in a manner similar to a keep-alive. Upon receipt of the Discard-Request the receiving endpoint simply tosses it away.

4.1.3.5 Link Termination

The third category of LCP packets involves link termination or the orderly shutdown of a connection. There are two packets used for link termination, Terminate-Request and Terminate-Ack. An endpoint that wants to close a connection will transmit a Terminate-Request packet. That packet has a code type field value of 05. Upon reception the opposite endpoint will respond with a Terminate-Ack which is identified by a code type field value of 06.

The originator of the Terminate-Request does not indicate the reason for the request. Instead, this packet simply informs the recipient to immediately terminate any ongoing operations. Thus, the originator could be shutting down a connection or responding to a higher layer request to terminate the session. The only thing the receiver knows is to acknowledge the request and terminate any ongoing operations.

Now that we have an appreciation of the three classes of LCP packets and their general use, we will conclude our discussion of PPP by discussing LCP configuration options.

4.1.3.6 Configuration Options

The goal of LCP configuration options is to provide a mechanism to support the negotiation of modifications to the default operating characteristics of a point-to-point link. The format of an LCP Configuration Option packet is shown in Figure 4.4. Note the Protocol field has a value of hex C021 to indicate it is an LCP packet. The Type field is 1 byte in length. Its value indicates the type of Configuration Options. Table 4.3 lists examples of Type field values for the Configuration Option and their meaning.

Returning our attention to the Configuration Option format shown in Figure 4.4, the Length field is also 1 byte. The value of this field indicates the length of the option to include Type, Length and subfields.

There are six defined configuration options. The Maximum-Receive-Unit option indicates the maximum amount of data an endpoint can receive. The Quality Protocol option is used to negotiate a specific protocol for

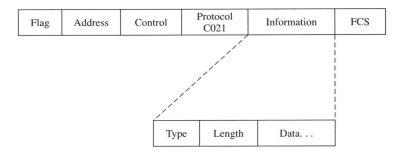

Figure 4.4 LCP Configuration Option format.

TABLE 4.3 Configuration Option Codes

Type Field Value	Meaning
0	RESERVED
1	Maximum-Receive-Unit
3	Authentication Protocol
4	Quality Protocol
5	Magic Number
7	Protocol-Field-Compression
8	Address-and-Control-Field-Compression

link quality monitoring. The Magic Number configuration option provides a method to detect looped-back links and other abnormal data link layer conditions. Here the term 'magic number' represents a unique bit stream, such as hex 1A2B3C4D. If the transmission of this bit stream results in its receipt at the point of origin, PPP will conclude that there is a loop on the circuit. The Protocol-Field-Compression option provides a means to negotiate the compression of the PPP protocol field, while the Address-and-Control Field Compression Configuration Option, as its name implies, provides a mechanism to negotiate the compression of the Address and Control fields. One additional configuration option, which we purposely saved for last, is the Authentication option, which we will now review.

Authentication
By default, authentication is optional and must be requested during the link establishment phase. Although authentication should occur as soon as possible after link establishment, it can occur concurrently with link quality determination. If authentication is required, the Configure-Request

will indicate it expects authentication from its peer. Under the PPP standard there is no requirement that authentication be full-duplex or that the same protocol be used in both directions.

Figure 4.5 illustrates the format of the Authentication Protocol Configuration Option. This protocol, as well as other types of optional data, are conveyed in the Data field of an LCP packet. In examining Figure 4.5 the Type field is 1 byte in length and is used to identify the configuration option. For the Authentication protocol the Type field value would be set to 3.

As previously noted, the Length field is 1 byte in length. Its value indicates the length of the option to include Type, Length and sub-fields. Thus, for Figure 4.5 the value of the Length field would be 4. The 2-byte Authentication Protocol field indicates the desired authentication protocol. For PAP this field would be set to a value of C023, while, if CHAP was desired as the authentication protocol, this field would be set to a value of C233.

4.1.4 Multilink PPP

An enhancement to PPP worth noting is Multilink PPP (MPPP). MPPP owes its development to the problems users encountered when transmitting PPP over ISDN. In certain situations a router would be configured to activate a second B channel, either when an ISDN call occurred or once a predefined traffic threshold was reached. Because packets could be transmitted over either channel and there is no guarantee that traffic across both channels would take an identical path through the ISDN network, it became possible for packets to arrive out of order at the destination router. If you were using PPP to transport time-sensitive or unreliable protocols, the arrival of out-of-sequence packets would obviously cause a problem. In addition, even when transporting a reliable protocol, such as TCP, a high number of out-of-sequence packets would result in a high level of retransmissions, which would significantly reduce throughput. A solution to the sequencing problem was obtained through the development of MPPP. MPPP adds a sequencing feature that enables data to be reassembled at layer 2. This eliminates the need for higher layer retransmission and in effect enhances link performance.

Figure 4.5 Authentication Protocol Configuration Option format.

MPPP is negotiated during the LCP setup process and results in the two ISDN B-channels operated as a logically aggregated data link.

4.2 Point-to-Point tunneling protocol

The Point-to-Point Tunneling Protocol (PPTP) was developed during the mid-1990s as a mechanism to tunnel PPP packets through an IP network. PPTP supports the secure transmission of data from a remote client through an IP network to a network server by creating a virtual private network between the two endpoints. As we will note later in this section, the remote client can be a PC operating PPTP software, an access concentrator, or even another network server, the latter resulting more accurately in server-to-server PPTP communications.

The original PPTP effort was based upon the efforts of a vendor consortium founded by Microsoft Corporation, U.S. Robotics (which later became a subsidiary of 3 Com Corporation) and ECI/Telemetrics. An Internet draft was developed and published in June 1996. The resulting RFC, RFC 2637 called 'Point-to-Point Tunneling Protocol (PPTP)' was published in July 1999.

4.2.1 Implementation models

When work commenced on PPTP, access to the Internet was primarily accomplished via dial modem access, with cable modems and Digital Subscriber Lines (DSLs) still many years from commercial availability. Due to this the developers of PPTP created a client−server architecture that could support three implementation models: client−server, access-concentrator−server, and server-to-server.

4.2.1.1 Client−Server Model

The client−server model occurs when a remote client, typically a PC operating PPTP client software that has a connection to the Internet tunnels via its IP network connection to a PPTP server. That server is typically located on the user's corporate network and can provide access to applications on the server or to other computers connected to the server that are located on the distant network. The upper portion of the network shown in Figure 4.6 labeled 'client−server' illustrates the PPTP client server model.

Advantages
Because the PPTP client creates a tunnel directly to the server, it does not require any special service to be performed by its Internet Service

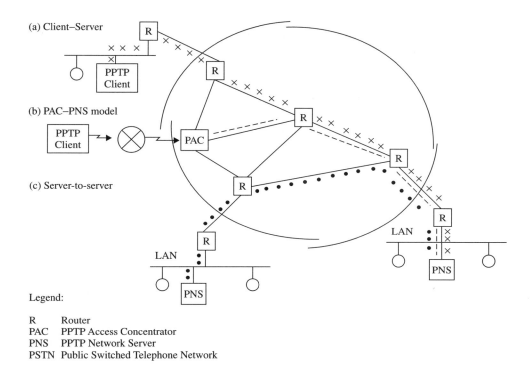

(a) Client–Server

PPTP
Client

(b) PAC–PNS model

PPTP
Client

PAC

(c) Server-to-server

R

R

R

R

R

R

LAN

PNS

R

LAN

PNS

Legend:

R   Router
PAC   PPTP Access Concentrator
PNS   PPTP Network Server
PSTN Public Switched Telephone Network

Figure 4.6   PPTP architectural models.

Provider (ISP). When we examine the next model we will note that that model depends upon equipment located at the ISP for the creation of a tunnel. A second advantage associated with the client–server model results from the fact that it can perform encryption and compression on an end-to-end basis. Due to this, the client–server and server–server models are more secure than the access concentrator–network server model.

Disadvantages

There are a few disadvantages with the client–server model, each related to the fact that each PC creates an individual session to the server. First, the client–server model requires an individual tunnel to be formed from source to destination, which can result in performance issues for both the ISP and the organization's network server. Second, because PPTP software must operate on the client, this means that legacy devices, such as Windows 3.1 machines for which PPTP is not available, cannot be used, although you should have retired such devices long ago.

4.2.1.2 Access Concentrator—Network Server Model

The second model supported by PTP is the access concentrator-network server model. This model, which is shown in the middle portion of Figure 4.6, results in the ISP taking on the role of the PPTP client. Here the ISP uses a PPTP compliant access concentrator to create a PPTP tunnel on behalf of its remote clients. The access concentrator is more formally referred to as PPTP Access concentrator, or PAC. In comparison, the network server is referred to as a PPTP network server or PNS. Thus, this model is commonly referred to as the PAC—PNS model.

Advantages

There are several advantages associated with the PAC—PNS model. First, because a PAC has the ability to multiplex several PPP sessions over a single tunnel, this means that each remote user does not need to set up a separate tunnel. Thus, several users accessing the same PNS can share the same PPTP control channel, which enhances the performance of this model over the individual client—server model with respect to the operation of the network server.

A second key advantage of the PAC—PNS model is the fact that it supports clients that cannot create a PPTP tunnel directly. This means that this model can support any operating system that has a PPP capability, since the client communicates with the access concentrator via PPP.

Disadvantages

The PAC—PNS model adds an intermediary in the form of an access concentrator between the client and network server. This means that the access concentrator must be operational for the client to create a tunnel to the server. Because the access concentrator can provide support to a group of clients, its failure is more serious than the failure of a single client.

Another disadvantage of the PAC—PNS model results in the fact that the secure tunnel is only from the access concentrator to the network server. Thus, encryption is not end-to-end, but only through the Internet.

4.2.1.3 Server-to-Server Model

The third PPTP architectural model is the server-to-server model. This model is illustrated in the lower portion of Figure 4.6. Under this model two servers are interconnected via a secure tunnel through the Internet or a private IP network. Although it is possible for any operating system to support this model, to the best of this author's knowledge Microsoft is the only vendor supporting a server-to-sever PPTP model.

Advantages

A key advantage of the PPTP server-to-server model lies in the fact that clients connecting to the server do not have to operate their own tunnels. Thus, this removes the burden of PPTP processing operations from remote clients. Another advantage associated with this model is the fact that if your organization is using Windows NT, Windows 2000 or a recently released Windows 2003-based server, you have this capability within the server operating system. Thus, you could implement PPTP to support workstations on two LANs via a server-to-server model without having to configure any remote clients.

Disadvantages

There are two key disadvantages associated with the server-to-server model. First, as previously discussed, it is limited to a Microsoft environment. Second, requiring all tunneling to include control information to flow between two servers could degrade their performance. This could adversely affect their ability to support applications on the server or access to other networked computers.

4.2.2 Networking functions

During the development of the PPTP standard considerable thought was given to the area of communications control. Because the access concentrator would interface the PSTN and any digital circuits, such as ISDN, the requirement for the network server to control modems and ISDN terminal adapters was not necessary. This situation enabled certain functions normally associated with a server, such as interfacing the PSTN and the direct control of external modems and ISDN terminal adapters to be uncoupled from the server. This also resulted in the access concentrator functioning as a modem or ISDN terminal adapter termination point and assuming responsibility for the logical termination of PPP Link Control Protocol (LCP) sessions. Because several remote clients can establish PPP sessions with the access concentrator, the latter becomes responsible for channel aggregation as well as the logical termination of PPP network control protocols.

At the network server, that device becomes responsible for terminating aggregated channels as well as for multiprotocol routing and bridging between its interfaces. The latter is required on the server regardless of the PPTP architectural model since the server needs the ability to direct traffic as well as receive traffic from other devices connected either directly or indirectly to the network.

The establishment of a PPTP connection to a network server either directly or via an access concentrator is a two-part process. First, the remote client will establish a PPP connection directly to the server or to the access concentrator. Once this occurs, a second call is established over the previously established PPP connection. The second call results in the flow of IP datagrams containing PPP packets, in effect encapsulating the packets within IP datagrams. The second call actually creates the VPN, with the connection between the remote client and server referred to as a tunnel. Thus, under PPTP, the VPN connection can be viewed as a PPP connection within another PPP connection. Of course, special control messages must flow between the access concentrator and network server to establish and control the operation of the tunnel through the IP network.

The movement of many networking functions from a traditional network server to the access concentrator provides a variety of networking advantages. For example, a client could connect to the access concentrator and contact different network servers while using a single IP address. Not only does this action reduce table space in the access concentrator, it also reduces the use of a scarce resource in the form of IP addresses. To better understand how this literal division of labor operates, let's briefly discuss some of the functions the access concentrator and network server perform, both individually and jointly.

4.2.2.1 Access Concentrator Functions

As indicated in Figure 4.6, the access concentrator terminates calls from the remote client. Those calls can occur via modem dial-up or via an ISDN connection. Thus, the access concentrator must be able to support both analog and digital connections as well as provide an interface to modems and ISDN terminal adapters.

To transport data from received calls into a serial data stream, the access concentrator must be capable of performing analog to digital and digital to analog conversion. In addition, the access concentrator will receive data in different transmission modes, requiring the need to support asynchronous to synchronous and synchronous to asynchronous conversions.

Because the access concentrator is responsible for contacting the network server, it needs to provide support for the PPP Link control Protocol to include managing each session established to the network server.

4.2.2.2 Network Server Functions

The network server acts as a logical termination point for various PPP control protocols. Because the remote client may require access to data on the server as well as data on other devices connected to the LAN, the server must be capable of supporting multiprotocol bridging and routing between its interfaces. In addition, when supporting ISDN the network server needs to be capable of aggregating channels into a bundle. This function is referred to as multilink management.

4.2.2.3 Shared Functions

There are certain functions both the access concentrator and network server share. For example, both devices must participate in PPP authentication protocols as well as in the use of data compression.

4.2.3 Establishing the PPTP Tunnel

There is a sequence of steps that must be performed to establish a tunnel between the PPTP client and the PPTP server. Those steps are based upon the manner by which the PPTP connection is established. That is, a remote client connected to a LAN can directly connect to a PPTP server using IP as a transport facility via their LAN connection, establishing a tunnel from client to server. In comparison, in a dial-up connection mode the client first uses the PPTP protocol to connect to the ISP. If they are using the services of a PPTP-compliant access concentrator, then a second connection using PPTP is made from the access concentrator to create a tunnel through the Internet to the PPTP server. Now that we have a general appreciation of the basic manner in which PPTP operates and how functions are delegated to different hardware devices, let's probe deeper into the protocol. In doing so, let's turn our attention to the packet encapsulation process.

4.2.4 PPTP encapsulated packets

There are two basic types of PPTP encapsulated packets: those that are used to transport control information and those that are used to convey data. Packets used to transport control connection information use a TCP connection to create, maintain and terminate the tunnel. The top portion of Figure 4.7 illustrates the format of the PPTP control connection packet. In comparison, data is transported as a payload in a PPP packet using a modified version of

(a) PPTP Control Connection Packet Format

Data Link Header	IP Header	TCP Header	PPTP Control Message	Data Link Trailer

(b) PPTP Tunneled Data Packet Format

Data Link Layer	IP Header	GRE Header	PPP Header	PPP Payload	Data Link Trailer

Figure 4.7   PPTP Packet formats.

the Generic Routing Encapsulation (GRE) protocol. This results in a series of headers being prefixed to the PPP header as illustrated in the lower portion of Figure 4.7. The actual payload transported can be an IP or IPX Datagram, an AppleTalk, or NetBEUI frame, or one of many obscure protocols. However, in the real word the primary protocol conveyed is IP. The payload can be encrypted or compressed or both. In actuality, PPTP inherits encryption and/or compression from PPP. When establishing the VPN PPTP authentication is based upon the same authentication mechanisms as supported by PPP, illustrating the close relationship between the two protocols. Now that we have a basic understanding of the two types of PPTP packets, let's turn our attention to obtaining specific information about each type of packet.

4.2.5 The PPTP control connection packet

PPTP Control Connection packets represent messages transmitted as TCP data on the control connection between either a remote client communicating directly with a network server or an access concentrator and network server. PPTP uses TCP port 1723 for the connection, which means that you need to ensure that TCP data on port 1723 can flow through organizational routers and firewalls.

Although the composition of the PPTP Control Message field shown in Figure 4.7a differs by message type, the first five subfields share commonality. Figure 4.8 illustrates the initial five fields within the PPTP Control Message field. Each PPTP control Message field begins with a 10-byte fixed header. That header contains five fields: a 2-byte Length field that indicates the length of the message, a 2-byte Message Type field that indicates the PPTP message type, a 4-byte Magic Cookie field, a 2-byte Control Message Type field and a 2-byte Reserved field. The Magic Cookie field is always set to the value hex 1A2B3C4D. This setting enables the receiver to synchronize itself to the TCP data stream.

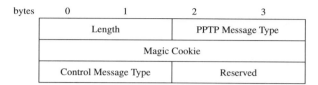

bytes	0	1	2	3

Figure 4.8 Initial five fields of the PPTP control message field.

There are two Control Connection message types defined by the PPTP Message Type field. A value of 1 in that field defines a Control Message, while a value of 2 defines a Management Message. Management messages are currently not defined. The PPTP Control Message field further defines the type of control message being conveyed. There are currently 15 defined control messages that can be grouped into four functional areas: control connection management, call management, error reporting and PPP session control. Table 4.4 lists the currently defined Control Message Type field values and their meanings.

TABLE 4.4 PPTP Control Messages

Functional Area/Control Message	Code Value
Control Connection Management	
Start-Control-Connection-Request	1
Start-Control-Connection-Reply	2
Stop-Control-Connection-Request	3
Stop-Control-Connection-Reply	4
Echo-Request	5
Echo-Reply	6
Call Management	
Outgoing-Call-Request	7
Outgoing-Call-Reply	8
Incoming-Call-Request	9
Incoming-Call-Reply	10
Incoming-Call-Connected	11
Call-Clear-Request	12
Call-Disconnect-Notify	13
Error Reporting	
WAN-Error-Notify	14
PPP Session Control	
Set-Link-Info	15

4.2.5.1 PPTP Control Messages

PPTP Control Messages are used to establish, maintain and provide for an orderly termination of the VPN connection. As indicated from a quick glance at Table 4.4, most control messages are in pairs – a request followed by a reply since control information flows between either the remote client or an access concentrator and network server. In examining the PPTP control messages we will focus our attention upon such messages by the functional area in which they can be categorized. Because the control message code values are nicely aligned with the functional areas where control messages reside, we will also review such messages by their message code value.

4.2.5.2 Control Connection Management

As indicated in Table 4.4, there are six PPTP control messages that can be considered as control connection management messages. The first two control messages, Start-Control-Connection-Request and Start-Control-Connection-Reply, are used to determine the Control Connection protocol version used.

Two additional control messages categorized as control connection management messages are Stop-Control-Connection-Request and Stop-Control-Connection-Reply. The function of this pair of messages is to provide for an orderly method for closing the PPTP connection, enabling the access concentrator and network server to process any pending messages.

The last pair of messages that fall into the control connection management messages category are Echo-Request and Echo-Reply. These messages function as keep-alives to inform the other party that in the absence of actual data the connection should be maintained. Now that we have an appreciation for the six control messages that fall into the control connection management message category, let's look at each one in detail.

Start-Control-Connection-Request
The Start-Control-Connection-Request control message is transmitted by the PPTP client to establish a connection between an access concentrator and network server. This message can be initiated by either device and must precede the issuance of any other control message.

Figure 4.9 illustrates the format of the Start-Control-Connection-Request Message. Remember, the fields shown in Figure 4.9 reside within the PPTP Control Message field previously shown in the top portion of Figure 4.7.

The first five fields shown in Figure 4.9 are common for all control messages and were previously described. Because the Start-Control-Connection-Request

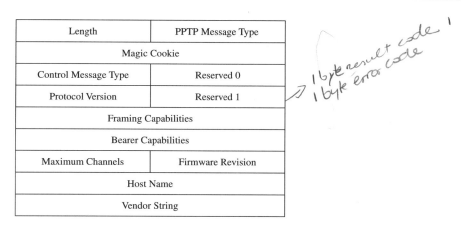

Length	PPTP Message Type
Magic Cookie	
Control Message Type	Reserved 0
Protocol Version	Reserved 1
Framing Capabilities	
Bearer Capabilities	
Maximum Channels	Firmware Revision
Host Name	
Vendor String	

(handwritten: 1 byte result code / 1 byte error code)

Figure 4.9   Start-Control-Connection-Request control message format.

represents a control message, the PPTP Message Type field is set to a value of 1 to identify the Control Message as a Start-Control-Connection-Request Message.

Continuing our tour of the fields shown in Figure 4.9, both Reserved fields and, in fact, all Reserved fields in any PPTP control message have their values set to 0. The Framing Capabilities field indicates the type of framing the message originator can provide. Defined settings are currently 1 for asynchronous framing and 2 for synchronous. The Bearer Capabilities field indicates the bearer capabilities that the message originator can provide. Currently defined settings are 1 for analog access and 2 for digital access support.

The Maximum Channels field indicates the number of PPP sessions the access concentrator can support. If this message is initiated by the network server, this field is set to a value of 0 and ignored by the access concentrator. The Firmware Revision field will contain the firmware revision number of the access concentrator or network server, with the entry dependent upon the hardware device that issued the control message. The last two fields in the message are the Host Name and Vendor String. The Host Name contains the DNS name of the access concentrator or network server that originated the message. In comparison, the Vendor String field contains a string that describes either the type of access concentrator or network server software being used, with the actual string dependent upon the device that issued the message.

Start-Control-Connection-Reply

The Start-Control-Connection-Reply control message is transmitted in response to a Start-Control-Connection-Request message. The format of this control message is similar to that shown in Figure 4.9; however, the second reserved field (Reserved 1) is subdivided into a 1-byte Result Code and a 1-byte Error Code field. The Result code indicates the result of the command channel establishment attempt. Table 4.5 lists valid Result Code field values.

The Error Code field is always set to a value of 0 unless an error exists, in which case the Result Code is set to a value of 2 and the Error Code field then has a non-zero value. When this situation occurs, the Error Code values correspond to General Error Code messages and are listed in Table 4.6.

Stop-Control-Connection-Request

The Stop-Control-Connection-Request is transmitted by one endpoint in an access concentrator to network server connection. The purpose of this control message is to inform the other party that the control connection should be closed. Either participant in a PPTP session can initiate this control message which not only closes the control connection but in addition clears all active user calls. The format of the Stop-Control-Connection-Request message is similar to that shown in Figure 4.9. However, the message is limited to 8 bytes after the Magic Cookie field, with the 2-byte Protocol Version field replaced

TABLE 4.5 Start-Control-Connection-Reply Result Codes

Code Value	Meaning
1	Successful channel establishment
2	General error, specific problem identified by Error Code value
3	Common channel exists
4	Requestor not authorized
5	Protocol version of requestor not supported

TABLE 4.6 General Error Codes

Error code	Meaning
0 (None)	No general error condition
1 (Not connected)	No control connection exists for the access controller–network server pair.
2 (Bad Format)	Length is wrong or Magic Cookie value is incorrect
3 (Bad Value)	One of the field values is out of range or reserved field value is non zero.
4 (No Resource)	Insufficient resources to handle this command.
5 (Bad Call ID)	The Call ID is invalid in this context.
6 (Access Controller Error)	A generic vendor-specific error occurred in the access controller.

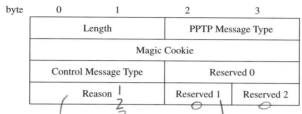

byte	0	1	2	3
	Length		PPTP Message Type	
	Magic Cookie			
	Control Message Type		Reserved 0	
	Reason		Reserved 1	Reserved 2

Figure 4.10   Stop-Control-Connection-Request control message format.

(handwritten annotations: "1 2", "Result code field", "Reply", "Error Code field", "1 2 3", "0", "0")

by a 1-byte Reason field, followed by a 1-byte and 2-byte Reserved field. This modified format is shown in Figure 4.10.

The Reason field indicates the reason why the Stop-Control-Connection-Request message was issued. Defined Reason codes include 1 to indicate a general request to clear the control connection, 2 to denote the peer cannot support the other end's version of the protocol and 3 to indicate the requester is being shut down. The Reserved fields continue to be set to 0.

Stop-Control-Connection-Reply

The Stop-Control-Connection-Reply control message is issued in response to the receipt of a Stop-Control-Connection-Request message. Because either participant in a PPTP session can issue the Stop-Control-Connection-Request control message, either participant can respond with the Stop-Control-Connection-Reply.

The format of the Stop-Control-Connection-Reply message is similar to the Stop-Control-Connection-Request message previously shown in Figure 4.10. The key difference between the two concerns the last 4 bytes in the message. Here the Reason field is replaced by a Result Code field, while the first reserved (Reserved 1) field is replaced by an Error Code field. The Result Code field is used to indicate the result of the attempt to close the connection, with a value of 1 used to indicate that the control connection is closed and a value of 2 used to indicate the control connection was not coded. If the value for the Result Code field is 2, then the Error Code field defines the reason for the control connection not closing. Those Error Code field values are general in nature and were listed in Table 4.6.

Echo-Request

The Echo-Request PPTP control message functions as a 'keep-alive' signal to keep the connection active in the absence of actual data transfer. Either participant in the PPTP session can issue an Echo-Request control message, with the other party responding with an Echo-Reply control message.

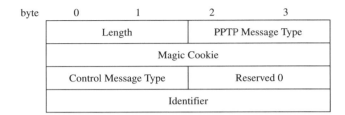

Figure 4.11   Echo-Request control message format.

Figure 4.11 illustrates the format of an Echo-Request Control message. Note that the Identifier field value is set by the sender of the Echo-Request control message. This value is returned in the corresponding Echo-Reply as a mechanism to match the two.

Echo-Reply
The Echo-Reply represents the PPTP control message response to an Echo-Request message. If this message is not received by the endpoint that issued the Echo-Request, the PPTP tunnel will be terminated.

The Echo-Reply message has four more bytes than the Echo-Request. These additional bytes include a 1-byte Result Code field and 1-byte Error Code field, followed by a 2-byte Reserved field to ensure the message ends on a fixed boundary. The Result Code values can be either 1 to indicate a valid reply or 2 to indicate the Echo-Request was not accepted. Concerning the latter, when an Echo-Request is not accepted, the reason for its rejections will be defined by the value in the Error Code field. Those Error Code field values are general in nature and were listed in Table 4.6. Both Echo-Request and Echo-Reply control messages are structured to keep a connection active and should not be confused with similar ICMP messages.

4.2.5.3 Call Management

A second category or grouping of PPTP control messages are those involved with call management. There are seven control messages that fall into this functional area that we will review in this section. Those messages were previously listed in Table 4.4 and have message code values of 7 through 13.

Outgoing-Call-Request
The purpose of the Outgoing-Call-Request control message is to create a PPTP tunnel from the network server to the access concentrator. Figure 4.12

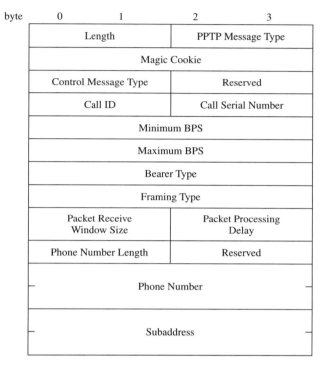

Figure 4.12   The Outgoing-Call-Request control message.

illustrates the format of the Outgoing-Call-Request control message. Fields within this message indicate the minimum (Minimum BPS) and maximum (Maximum BPS) data rate in bps that defines the lowest and highest acceptable line speed for the session. Two additional fields worth noting are a Call ID and Call Serial Number field. The Call ID field represents a unique identifier assigned by the network server to the session and provides a mechanism to multiplex and demultiplex data flowing between the access concentrator and network server. In comparison, the Call Serial Number, which is also assigned by the network server, is used to identify the session in a logged session database of information.

The Bearer Type field contains a value that indicates the bearer capability required for the outgoing call. Currently defined values include 1 for a call to be placed on an analog channel, 2 for a call to be placed on a digital channel and 3 for a call that can be placed on any type of channel.

The Framing Type field contains a value that indicates the type of framing to be used for the outgoing call. Values defined include 1 for asynchronous framing, 2 for using synchronous framing, and 3 if the call can use either type of framing.

Similar to TCP, the Packet Receive Window Size field indicates the number of received data packets the network server will buffer for the session. By altering the value in this field the receiver in effect obtains a flow control capability.

The Packet Processing Delay field contains a value specified in tenths of a second. That value indicates the packet processing delay that can be expected on data sent to the server from the access concentrator. The Phone Number Length field indicates the valid number of digits in the Phone Number field, while the Subaddress field contains additional dialing information.

Outgoing-Call-Reply

The response to an Outgoing-Call-Request control message is an Outgoing-Call-Reply. The Outgoing-Call-Reply flows from the access concentrator to the network server, providing the server with information indicating the result of the outgoing call attempt as well as information the server can use to regulate the flow of data to the access concentrator.

Figure 4.13 illustrates the format of the Outgoing-Call-Reply control message. The first six fields of the Outgoing-Call-Reply are the same as the Outgoing-Call-Request control message. However, after the sixth field the two formats diverge, with the Reply message containing a Participant Call ID in place of the Call Serial Number. The value in this field is set to the value received in the Call ID field of the corresponding Outgoing-Call-Request message, enabling the Reply to match the Request. Instead of the two BPS and two Framing fields, the Reply message has a 1-byte Result Code and 1-byte Error Code fields, followed by a 2-byte Cause field and a 4-byte Connect Speed field. Of key importance is the Result Code field whose values indicate the result of the Outgoing-Call-Request attempt transmitted by the network server. Table 4.7 lists presently defined Result Code field values and their meanings.

Similar to other control messages, a value of 2 in the Result Code field indicates a general error. That error is then defined in the Error Code field whose values were previously listed in Table 4.6. The Cause field provides additional information about a failure, with the value of this field dependent upon the type of call attempted. For example, an ISDN call attempt failure will result in the Cause Code field containing an ISDN Q.931 Cause Code value.

The Connect Speed field indicates the connection speed in BPS. The Receive Packet Window Size field indicates the number of received data packets the access concentrator can buffer for the session. The Packet Processing Delay field functions as previously described, but the delay measurement is on data

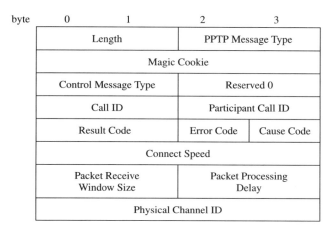

byte   0   1   2   3

Length	PPTP Message Type
Magic Cookie	
Control Message Type	Reserved 0
Call ID	Participant Call ID
Result Code	Error Code	Cause Code
Connect Speed	
Packet Receive Window Size	Packet Processing Delay
Physical Channel ID	

Figure 4.13   Outgoing-Call-Reply control message format.

TABLE 4.7   Outgoing-Call-Reply Result Code values

Result Code Field Value	Meaning
1 (Connected)	Call established with no errors
2 (General Error)	Outgoing call not established, reason denoted in Error Code
3 (No Carrier)	Outgoing call failed due to no carrier detected
4 (Busy)	Outgoing call failed due to detection of a busy signal
5 (No Dial Tone)	Outgoing call failed due to lack of dial tone
6 (Time Out)	Outgoing call not established with time allocated by access controller
7 (Do Not Accept)	Outgoing call administratively prohibited

sent from the access concentrator to the network server. The last field, Physical Channel ID, is set by the access concentrator to indicate the physical channel number used to place the call and represents a vendor-specific value.

Incoming-Call-Request

As previously noted, the Outgoing-Call-Request sets up a tunnel from the network server to the access concentrator. In comparison, a sequence of three control messages, of which the Incoming-Call-Request is the first, is required to notify the server that an inbound call needs to be established from the access concentrator. Thus, you can consider the Incoming-Call-Request control message as the first in a 'three-way message handshake' used by PPTP for establishing incoming calls.

Fields in the Incoming-Call-Request control message will identify if the call has arrived on an analog or digital channel (Call Bearer Type field) and the physical channel of the access controller the call arrived on (Physical Channel ID field). They will also identify the telephone number that was dialed by the caller (Dialed Number field), the number from which the call was placed (Dialing Number field) and additional dialing information that will be contained in the Subaddress field. Figure 4.14 illustrates the format of the Incoming-Call-Request control message.

Incoming-Call-Reply

The Incoming-Call-Reply control message can be considered as the second in the three-way handshake used by PPTP for establishing incoming calls. This control message, whose format is shown in Figure 4.15, is sent by the network server to the access concentrator in response to an Incoming-Call-Request message. Included in the Incoming-Call-Reply control message is a Result Code field whose value indicates the result of the Incoming-Call-Request attempt. Defined field values include 1 to indicate the access concentrator should answer the incoming call, 2 to indicate the call should not be established, and 3 to tell the access concentrator not to accept the incoming call. Similar to other control messages the Error Code field is set to a value of 0 unless a general error condition occurred and is noted by a Result

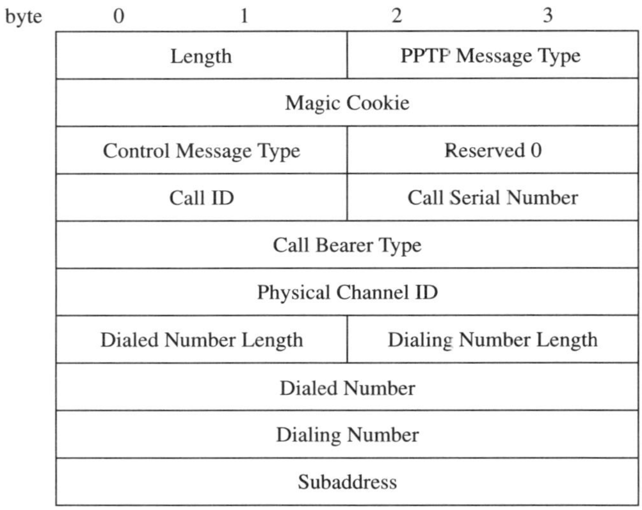

byte	0	1	2	3
Length		PPTP Message Type		
Magic Cookie				
Control Message Type		Reserved 0		
Call ID		Call Serial Number		
Call Bearer Type				
Physical Channel ID				
Dialed Number Length		Dialing Number Length		
Dialed Number				
Dialing Number				
Subaddress				

Figure 4.14   Incoming-Call-Request control message format.

byte	0	1	2	3

Length	PPTP Message Type
Magic Cookie	
Control Message Type Call ID	Reserved 0 Participant Call ID
Result Code	Error Code
Receive Packet Window Size	Packet Transmit Delay
Reserved 1	

Figure 4.15 Incoming-Call-Reply control message format.

Code field value of 2. When this situation occurs, the Error Code values take on the meanings previously listed in Table 4.6.

Incoming-Call-Connected

The Incoming-Call-Connected represents the third control message in the three-way handshake used for establishing incoming calls. This message is transmitted by the access controller to the network server in response to a received Incoming-Call-Reply message. Information included in this message includes access control parameters to be used for the call as well as information that enables the network server to regulate the flow of data to the access concentrator. Figure 4.16 illustrates the format of the Incoming-Call-Connected message.

Length	PPTP Message Type
Magic Cookie	
Control Message Type	Reserved 0
Participant Call ID	Reserved 1
Connect Speed	
Receive Packet Window Size	Packet Transmit Delay
Framing Type	

Figure 4.16 Incoming-Call-Connected message format.

In examining the fields in the Incoming-Call-Connected control message shown in Figure 4.16, the first five function as previously described. The Participant Call ID field is set to the value received in the Call ID field of the corresponding Incoming-Call-Reply message. This value is used by the network server to match the Incoming-Call-Connected message with the Incoming-Call-Reply it issued. The remaining fields in Figure 4.16 function as previously noted. Figure 4.17 shows the relationship between the access concentrator and network server with respect to the three control messages in the three-way handshake.

Call-Clear-Request

The purpose of the Call-Clear-Request control message is to provide the network server with the ability to initiate a call disconnect. The server sends this message to the access concentrator regardless of whether the call being cleared is an incoming or outgoing call.

Figure 4.18 illustrates the format of the Call-Clear-Request control message. Other than the first five fields whose values and use were previously described, the only field worthy of mention is the Call ID field. The network server assigns the value in the Call ID field for this control message. The reason for using a Call ID instead of a Participant Call ID results from the fact that the latter may not be known to the network server if the call is aborted during call establishment.

Call-Disconnect-Notify

The Call-Disconnect-Notify Control message is transmitted by the access controller to the network server. The purpose of this control message is to

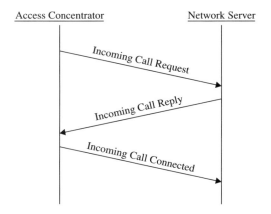

Figure 4.17   The incoming call three-way handshake.

byte   0   1   2   3

Length	PPTP Message Type
Magic Cookie	
Control Message Type	Reserved 0
Call ID	Reserved 1

Figure 4.18   Call-Clear-Request control message format.

inform the network server that a call was disconnected, as well as the reason for the disconnect. Although normally the Call-Disconnect-Notify message is issued in response to the receipt of a Call-Clear-Request control message, the access controller will also issue this message in the event a call is disconnected.

Figure 4.19 illustrates the format of the Call-Disconnect-Notify control message. Of particular interest are the Call ID, Result Code and Call Statistics fields. The Call ID field value represents the value assigned by the access concentrator to the call. This value is used instead of a Participant Call ID because the latter may not be known to the network server if the call had to be aborted during call establishment.

The Result Code field value indicates the reason for the disconnect. Current values include 1 when a call is lost due to loss of a carrier, 2 for a general error condition in which the Error Code defines the reason according to the values in Table 4.6, 3 for a call disconnected for administrative purposes and 4 for a call disconnected due to the receipt of a Call-Clear-Request control message. The Call Statistics field contains an ASCII string containing vendor-specific call statistics information that can be used for diagnostic purposes.

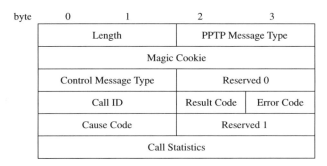

byte   0   1   2   3

Length	PPTP Message Type	
Magic Cookie		
Control Message Type	Reserved 0	
Call ID	Result Code	Error Code
Cause Code	Reserved 1	
Call Statistics		

Figure 4.19   Call-Disconnect-Notify control message format.

4.2.5.4 Error Reporting

A third category of PPTP control message is error reporting. There is currently only one control message that falls within this category: the WAN-Error-Notify message, which is described next.

WAN-Error-Notify

The WAN-Error-Notify control message is transmitted from the access concentrator to the network server. The purpose of this message is to enable the access concentrator to indicate to the network server error conditions that occurred on its interface that supports PPP. Included in this message are several counters that function cumulatively and represent a mechanism for determining the health of the PPP connection. This message should only be sent when an error occurs and not more than once every 60 seconds.

Figure 4.20 illustrates the format of the WAN-Error-Notify Control message. The first five fields are used as previously described. The Participant Call ID field contains the Call ID assigned by the network server to the call. Of particular interest are the error and overruns fields, whose inclusion results in the name of this control message.

The CRC Errors field indicates the number of PPP frames received with CRC errors since the session was established. By counting frames and examining the CRC Errors field value, it becomes relatively easy to determine the frame

byte	0	1	2	3
Length		PPTP Message Type		
Magic Cookie				
Control Message Type		Reserved 0		
Participant Call ID		Reserved 1		
CRC Errors				
Framing Errors				
Hardware Errors				
Buffer Overruns				
Timeout Errors				
Alignment Errors				

Figure 4.20 WAN-Error-Notify control message format.

error rate. The next field, Framing Errors, indicates the number of improperly framed PPP packets received. This field is followed by a Hardware Errors and a Buffer Overruns field, which indicate the number of receive buffer overruns and number of buffer overruns detected since the session was established. The last two fields in the message, Timeout Errors and Alignment Errors, indicate the number of timeouts and alignment errors since the call was established. When a new call is established, the previously described counters are reset.

4.2.5.5 PPP Session Control

The last category of PPTP control messages is PPP Session Control. There is currently one defined control message that falls into this category: the Set-Link-Info control message.

Set-Link-Info
The Set-Link-Info control message is transmitted by the access concentrator to the network server as a mechanism to define PPP-required link options. Because such options can change during the life of a connection, the access concentrator will update its internal call information dynamically and perform PPP negotiation on an active PPP session.

Figure 4.21 illustrates the format of the Set-Link-Info control message. The Asynchronous Control Character Map (ACCM) fields are bit map fields that enable (set to 1) or disable (set to 0) the use of character escapes for asynchronous links for the 32 ASCII control characters. By default the Send ACM and Receive ACM values used by the client until the Set-Link-Info message is received is hex FFFFFFFF.

4.2.6 Control connection protocol operation

As previously noted earlier in this section, PPTP control connection messages use TCP as the transport mechanism. The use of TCP results in a reliable

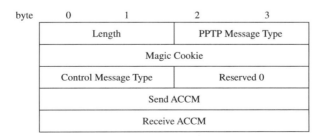

Figure 4.21 Set-Link-Info control message format.

transport mechanism that negotiates the need for PPTP to manage the connection by performing error detection and recovery operations. However, because the TCP connection can close at any time, this means that PPTP must be capable of being able to compensate for this situation. Thus, PPTP connection management requires the ability to perform error recovery procedures to include the appropriate logging of invalid or malformed messages to facilitate problem resolution.

A second area where the use of TCP as a transport mechanism can create problems concerns performance over high latency networks that may have a high packet loss rate. Because TCP is a reliable protocol, the error detection and correction can further complicate latency. In addition, the relatively lengthy header (in comparison to UDP) further complicates performance. Fortunately, Microsoft issued updates to its PPTP support for both client and server that improve performance when PPTP is operating over high latency networks and/or networks with relatively high levels of packet loss. Thus, if you are using PPTP it is highly recommended that you ensure you are working with the latest version of PPTP. Now that we have an appreciation of the operation of PPTP control messages we will conclude this section by turning our attention to data tunneling.

4.2.7 PPTP data tunneling

When we began our examination of PPTP we noted there were two types of packet formats: one for control messages and a second for the transportation of data. In concluding our examination of PPTP we will turn our attention to PPTP data tunneling.

In examining the lower portion of Figure 4.7 we can note that user data is transmitted through a series of encapsulations performed on the PPP payload. First, the initial PPP payload is encrypted and encapsulated with a PPP header prefixed to the payload to form a PPP frame. However, unlike a standalone PPP frame, this frame does not include an HDLC flag nor control characters or a CRC field. Next, the PPP frame is encapsulated through the use of a modified GRE header whose use facilitates the routing of encapsulated data transmitted over an IP network. In fact, GRE represents a client protocol of IP, which is identified as following the IP header by the setting of the Protocol field in the IP header to a value of 47. Because the GRE header is used to provide a general purpose method for routing encapsulated data, let's look at this header.

4.2.7.1 The GRE Header

The Generic Routing Encapsulation (GRE) header used in PPTP represents a slight enhancement to GRE. GRE, which was developed in 1994, was one of the earliest tunneling protocols and was used as an encapsulation method for several pioneering VPN technologies, such as those from Ascend Corporation (now part of Lucent) and Bay Networks (now part of Nortel). The term 'Generic' results from the fact that the developers of GRE were tasked with supporting a variety of different payload types that operated on networks during the mid-1990s. To accomplish this, GRE was structured to include a Protocol Type field within its header that enabled different packet types to be encapsulated. Another interesting aspect of GRE is the use of a 4-byte Key field within its header. This field enables the source of a GRE packet to be identified and/or authenticated. This makes GRE well suited as a VPN protocol.

Because the delivery mechanism for GRE is open to the user, it can be transported by different network layer protocols. However, due to the growth in the use of the Internet and private IP-based networks to the detriment of other network protocols, it should come as no surprise that IP is by far the most common protocol used for the transmission of GRE packets.

Figure 4.22 illustrates the format of the enhanced GRE header while the legend provides a summary of the fields in the header. For PPTP, the GRE header was modified in three ways. First, an acknowledgement bit (A bit) was added. When set, this indicates that a 32-bit Acknowledgement Number field is present and that this field's value is significant. Thus, the presence of an Acknowledgement Number field represents a second modification to the GRE header. The third modification resulted in the splitting of the 4-byte Key field and its replacement with a 16-bit Payload Length field and a 16-bit Call ID field.

In examining the description of the fields in Figure 4.22, a few words of clarification are in order concerning the relationship of the Sequence Number and Acknowledgement Number fields. The Sequence Number is present if bit 3 in the header (s bit) is set to a value of 1. Similarly, the Acknowledgement Number field will be present if bit 8, the A bit, is set to 1.

The purpose of sequence numbers is to provide a mechanism to uniquely identify each packet. Sequence numbers are initialized to zero at session startup and payloads for a given user session that has the S-bit set are assigned the next consecutive sequence number for the session.

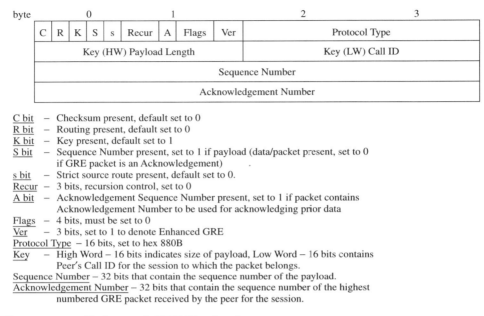

byte

C	R	K	S	s	Recur	A	Flags	Ver	Protocol Type
Key (HW) Payload Length									Key (LW) Call ID
Sequence Number									
Acknowledgement Number									

C bit   –   Checksum present, default set to 0
R bit   –   Routing present, default set to 0
K bit   –   Key present, default set to 1
S bit   –   Sequence Number present, set to 1 if payload (data/packet present, set to 0
            if GRE packet is an Acknowledgement)
s bit   –   Strict source route present, default set to 0.
Recur –   3 bits, recursion control, set to 0
A bit   –   Acknowledgement Sequence Number present, set to 1 if packet contains
            Acknowledgement Number to be used for acknowledging prior data
Flags   –   4 bits, must be set to 0
Ver   –   3 bits, set to 1 to denote Enhanced GRE
Protocol Type – 16 bits, set to hex 880B
Key   –   High Word – 16 bits indicates size of payload, Low Word – 16 bits contains
            Peer's Call ID for the session to which the packet belongs.
Sequence Number – 32 bits that contain the sequence number of the payload.
Acknowledgement Number – 32 bits that contain the sequence number of the highest
            numbered GRE packet received by the peer for the session.

Figure 4.22   Enhanced GRE Header format.

4.2.7.2 Data-Link Encapsulation

In concluding our discussion of PPTP data tunneling a few words are in
order concerning the manner by which the IP datagram, shown in the lower
portion of Figure 4.7, operates. When the IP datagrams flows on a LAN, the
header and trailer will represent the data-link layer LAN header and trailer.
For example, when the IP datagram flows on an Ethernet LAN, the header
will be the Ethernet Preamble field followed by the Ethernet Destination
Address, Source Address and Type or Length fields. Because the IP datagram
is carried in the Ethernet data or payload field, this also means that the
trailer is the Ethernet trailer, which is the 4-byte FCS field. Similarly, when
IP datagrams are transmitted over a point-to-point WAN via a modem or
ISDN connection, each datagram is encapsulated with a PPP header and
trailer. Upon receipt of tunneled data a 'de-capsulation' process occurs.
First, the data-link headers and trailers are removed. Next, the IP header,
GRE header and PPP headers are removed, leaving the PPP payload. If
encryption and/or compression has already occurred, then decryption and/or
decompression must take place, resulting in the payload being reconstructed
into its original form.

4.3 Layer Two Forwarding

A second tunneling protocol that deserves mention is the Layer 2 Forwarding (L2F) protocol. Although this protocol is very similar to PPTP in that it encapsulates PPP as a mechanism to create remote access VPNs over an IP network, L2F can be considered as being replaced by a more modern protocol referred to as Layer 2 Tunneling Protocol (L2TP). While we will discuss L2TP in the next section of this chapter, a few words about L2F are in order to better appreciate L2TP.

4.3.1 Evolution

At approximately the same time PPTP was being developed a second consortium of vendors including Cisco Systems, Northern Telecom (Nortel) and Shiva Corporation were developing a different method to encapsulate PPP as a mechanism to create VPN tunnels from remote access concentrator equipment. This second method, which dates to 1996, was published as RFC 2341 in May 1998. Titled Cisco Layer Two Forwarding (Protocol) 'L2F,' this RFC specified the tunneling of link data to include both PPP and SLIP frames. Unlike PPTP, L2F uses UDP as the transport protocol. In addition, L2F was limited to providing a tunnel from the access concentrator to the network server, unlike PPP that supports three architectural models. Recognizing that there were many positive aspects associated with both PPTP and L2F, the Internet Engineering Task Force (IETF) was a driving factor for both industry consortiums to work together and combine the best features of each independently developed tunneling protocol into a new standard referred to as L2TP. While L2TP represents the best of both worlds, PPTP continues to be a viable tunneling protocol due to the backing of Microsoft and the large core of Windows-based computers that support this technology. In contrast, L2F gradually faded away from use.

4.3.2 Operation

With Cisco Systems being a leading router manufacturer, it should come as no surprise that L2F tunnel mapping flows from a network access server (NAS), which is the L2F name for an access concentrator in the PPTP world, to the user's home gateway (HGW). A remote user accesses the NAS, which creates a tunnel to the home gateway. Once a tunnel is created, a field within the L2F header, referred to as the Multiplex ID, is assigned an unused value to denote the fact that a new dial-up session exists. The home gateway can either accept or reject the new session, with a rejection including a reason indicator, followed by a termination of the call by the NAS.

During the initial PPP call from the remote user to the NAS the ISP will use CHAP or PAP as a mechanism to authenticate the user. This authentication process may only represent what is referred to as 'partial authentication,' since the initial setup notification also allows the home gateway to authenticate the remote user and then decide to accept or reject the connection. To avoid an additional cycle of authentication, the home gateway can elect to use CHAP or PAP data previously submitted by the remote user.

The previously described method of authentication under L2F represents a key difference between this protocol and PPTP. Under L2F, user authentication at the remote end is performed by the home gateway and not by the network access server. This should come as no surprise since a major proponent of L2F is Cisco, while the major backer of PPTP is Microsoft, a vendor better known for its operating system software than its routing software.

4.3.3 The L2F packet format

The top portion of Figure 4.23 illustrates the format of an L2F packet. Note that the payload can be either SLIP or PPP, while the Checksum field is optional. The lower portion of Figure 4.23 illustrates the fields within the L2F header. By examining those fields we can obtain an appreciation of the capability and functionality of this pioneering tunneling protocol.

4.3.3.1 Flag Bits

The first byte in the L2F header contains four defined flag bits, labeled F, K, P and S. The F-bit is a Payload Offset identifier, which when set indicates an

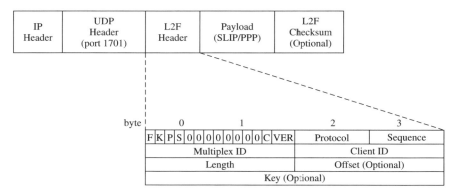

Figure 4.23 The L2F Packet format.

offset to the payload exists. Otherwise, a bit setting of 0 indicates no offset is present. The K-bit flag is set if the Key field is present in the header. Otherwise, a setting of 0 indicates that field is not present. The P-bit when set indicates the packet has priority and should be processed ahead of other packets that do not have this bit set. In the fifth bit position of the second byte is another one-bit flag. Here the C-bit when set indicates that the optional Checksum field is present.

4.3.3.2 Version Field

Similar to most protocols, L2F includes a Version (VER) field. The 3-bit field indicates the major version of L2F creating the packet.

4.3.3.3 Protocol Field

The 8-bit Protocol field indicates the protocol carried within the L2F packet. Defined values are hex 00 for illegal, hex 01 for L2F management packets, hex 01 for PPP tunneled within L2F, and hex 03 for SLIP tunneled within L2F.

4.3.3.4 Sequence Field

The 8-bit Sequence field is present only if the S-bit is set to 1. The Sequence field is mandatory for all L2F management packets. When this field is used, sequence numbering commences at 0 for the first packet and the sender increments the field value by 1. The receiver checks the value in this field and will silently discard a packet whose Sequence field value is less than or equal to the last received value.

4.3.3.5 Multiplex ID Field

The Multiplex ID field is used to identify a particular connection within a tunnel. Each new connection between the NAS and home gateway is assigned an unused MID value. The value of 0 is special, used to communicate the state of the tunnel.

4.3.3.6 Client ID Field

The Client ID (CLID) field provides a mechanism to assist endpoints in demultiplexing tunnels. This is especially important as it enables packets to be transmitted in an interleaved manner, with the CLID enabling demultiplexing to occur.

4.3.3.7 Length Field

The Length field denotes the size in bytes of the entire packet other than the optional checksum. If the receiver denotes the arrival of a packet whose length is less than the value of this field, it will silently discard the packet.

4.3.3.8 Checksum Field

This 16-bit field is optional, being present when the C-bit field is set. When this occurs, the checksum represents a 16-bit CRC computed over the packet from the first byte of L2F flag through the last byte of payload data.

4.3.3.9 Payload Offset Field

The Offset field is present when the F-bit in the first byte of the header is set. The purpose of this field is to indicate the number of bytes past the L2F header at which the payload data commences.

4.3.3.10 Key Field

This optional field is present when the K-bit is set. This field functions as a mechanism to resist attacks based on spoofing. The value of this field is obtained by taking the 128-bit authentication response from a peer as four adjacent 32-bit words and XORing them together.

4.3.4 Tunnel operations

Under L2F, when a connection is initiated between the NAS and Home Gateway, the endpoints first communicate using an MID 0 value. Doing so enables each endpoint to verify the presence of L2F on the opposite end as well as to permit any required authentication. When the L2F tunnel is being established, the first packet provides configuration data similar to a PPTP exchange. This packet will have a Protocol field value of 1 (L2F Management), while the Key, MID, CLID and Sequence fields are each set to a value of 0. The Key field is set to 0 as the tunnel is in the process of being initiated. The Sequence number begins at 0, while the MID field is set to 0 to reflect the establishment of the tunnel. Because the NAS at this point in time has not received a configuration packet from the gateway, the CLID field is also set to a value of 0.

Through a series of L2F management frames exchanged between the HGW and the NAS, the HGW is provided with several items of information. Those items of information include the name of the NAS and a challenge random number that the HGW will use in authenticating itself as a valid tunnel endpoint. The home gateway will respond to the NAS with a configuration in the form of another L2F management frame. This frame will have a sequence field value of 0 as each side maintains its own sequence numbering. The MID will be 0 to reflect tunnel establishment. A CLIP value will be set while the Key field value will continue to be 0 during tunnel establishment. Another series of configuration management frames will contain the HGW's name, its own random number challenge and its own assigned CLID for the NAS to place in the CLID field of future packets. The NAS will respond to the configuration with its Key field set to reflect the shared secret. This 32-bit Key represents the MD5 digest resulting from encrypting the shared secret, which will be used by both parties for the life of the tunnel.

4.3.5 Management messages

Similar to PPTP, L2F has a structured series of management messages. Those messages include a message used for requesting the configuration of the opposite endpoint, configuration messages that contain the name of the peer, its random number to be used as a challenge, and the assigned CLID. A series of L2F 'Open' management messages are used to establish a tunnel and the presence of a new dial-in client. Such 'Open' messages are acknowledged with several types of L2F 'Ack' messages. Rounding out the series of L2F management messages are another pair of 'Open' messages that convey the type of authentication used and the ID associated with authentication, an Echo/Echo Response pair of messages, and three 'Close' messages. Concerning the latter, one message is used to request a disconnect, while the other two indicate the reason for closure and convey an ASCII string description associated with the closure. Because L2F is no longer a viable tunneling protocol, having been either supplemented or displaced by PPTP, L2TP and IPSec, we will not further discuss this protocol. Instead, we will now turn our attention to L2TP, which represents a combination of features associated with PPTP and L2F.

4.4 Layer Two Tunneling Protocol

The Layer Two Tunneling Protocol (L2TP) can be considered to represent a combination of features incorporated into PPTP and L2F. In fact, as discussed earlier in this chapter, L2TP owes its existence to the development of those two competitive tunneling protocols. Because PPTP and L2F were incompatible with one another, their existence resulted in a considerable degree of confusion in both the vendor and end-user community. In an attempt to negate this confusion as well as to obtain the advantages associated with a common standard, the IETF directed the proponents of PPTP and L2F to combine their efforts in a common tunneling protocol. The result of this new effort was L2TP, which was published as RFC 2661 in August 1999.

4.4.1 Overview

Similar to both PPTP and L2F, L2TP encapsulates PPP frames. However, the encapsulation process results in the use of UDP as the transport mechanism for both data and management messages, leaving the performance of error detection and correction to higher layers in the protocol stack. Although L2TP currently is restricted for use over IP networks, it has the flexibility to transport tunneled data over X.25 packet networks, Frame Relay networks and even ATM.

4.4.2 Architectural models

L2TP supports two architectural models in comparison to the three models supported by PPTP and the single model supported by L2F. The first model supported by L2TP consists of communications between a concentrator and network server, while the second model consists of a remote client directly communicating with a network server. Unfortunately, L2TP uses a new series of nomenclature to describe the access concentrator and network server as illustrated in Figure 4.24. Under the concentrator to server model, L2TP refers to the access concentrator as an L2TP access concentrator, or LAC. Similarly, the network server is now referred to as an L2TP network server, or LNS.

Like PPTP, an L2TP session commences with a remote system initiating a PPP connection to a LAN. The L2TP access concentrator is responsible for terminating analog modem as well as digital ISDN calls. In addition, the L2TP architecture extends the PPPP and L2F models by allowing the LAC endpoints to reside on an Asynchronous Digital Subscriber Line (ADSL) access module or a similar cable modem device. The LAC may authenticate the user, subsequently using the same authentication method during the creation of an L2TP tunnel as in the initial PPP connection. If a remote user runs L2TP directly they can bypass the LAC and communicate directly to the LNS as shown by the tunnel in Figure 4.24. When this architectural model is used, authentication is provided by the LNS, either directly or by the LNS working in conjunction with another device, such as a RADIUS server.

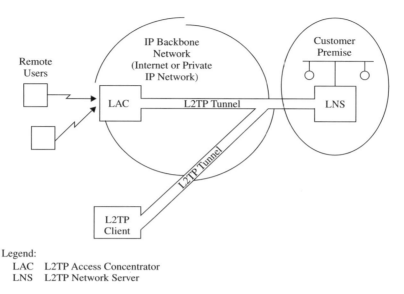

Figure 4.24   L2TP architectural models.

4.4.2.1 Encryption

One of the key differences between L2TP and its predecessor tunneling protocols is in the area of encryption. Encryption for L2TP connections occurs via the use of IPSec Encapsulated Security Payload (ESP). Because encryption is separated from this tunneling protocol it is possible to create a tunnel under certain operating systems, such as Windows 2000 that are not encrypted by IPSec. When you do this, you are not actually establishing a VPN connection to the LNS since the data encapsulated through the use of L2TP is not encrypted. Other than using a 'clear' tunnel for facilitating diagnostic testing you should always ensure L2TP tunnels are configured to use IPSec encryption. Concerning IPSec, this topic will be covered in detail in Chapter 5.

4.4.3 The L2TP packet format

The top portion of Figure 4.25 illustrates the format of an L2TP packet. Note that the L2TP header is used to encapsulate a PPP header and its payload, with UDP providing a transport mechanism. UDP port 1701 is used to transport L2TP, which means any tunnel that traverses organizational routers or firewalls must be configured to permit the flow of data on that UDP port. Also note that this basic packet format does not provide for encryption. To encrypt tunneled data requires the use of IPSec, which would modify the packet format by the insertion of an IPSec header between the IP header and the UDP header. The lower portion of Figure 4.25 illustrates the breakout of the L2TP header, which you will note has more similarities to L2F than PPTP. To obtain an appreciation of the manner by which L2TP operates, we will turn our attention to examining the fields in its header.

4.4.3.1 The L2TP Header

The format of the L2TP header borrows several concepts from L2F, such as the use of flag bits whose settings indicate if an optional field in the header is present. Because the first byte in the header begins with eight flag bits we will commence our discussion with those flag bits.

Flag bits
Each of the first eight bits in the L2TP header has a defined meaning and can be collectively referred to as flag bits since their settings indicate one of two conditions. The first bit is the Type (T) bit, whose setting indicates the type

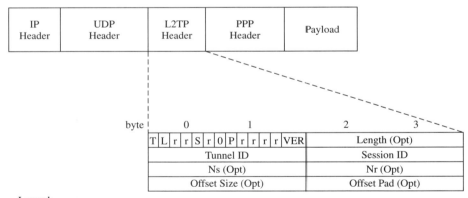

Legend:
T Type bit
L Length bit
r Reserved bit
S Sequence bit
O Offset bit
P Priority bit
Ver Version
Opt Optional

Figure 4.25 L2TP packet and header formats.

of message. A bit setting of 0 indicates a data message, while a bit setting of 1 indicates a control message. The use of the Type bit enables a common format to be used for the transmission of data and control messages.

The second bit in the L2TP header is the Length (L) bit. When this bit is set to 1, it indicates that the optional length field is present. The next two bits, which are labeled 'r' in the lower portion of Figure 4.25, are reserved for future use. That pair of reserved bits are followed by a Sequence (S) bit. When the S bit is set to a value of 1, the optional Ns and Nr fields are present.

The Offset (0) bit is set to 1 to indicate the presence of the Offset Size field, while the Priority (P) bit is set to 1 to indicate that this packet should receive preferential treatment for local queuing and transmission. For control messages the L-bit and S-bit must be set to 1 while the 0 bit must be set to 0.

Version field
The last four bits in the second byte of the L2TP header contain the version number. The value of this field must be 2 for an L2TP message header since 1 is reserved to permit the detection of L2F packets.

Length field
The 16-bit Length field is optional and included in the header when the L bit is set. The value in this field indicates the length of the message in bytes.

Tunnel ID field
The Tunnel ID field represents an identifier for the control connection. Each L2TP tunnel is identified by a 16-bit value that has local significance, resulting in the same tunnel having different tunnel IDs assigned by each endpoint. The Tunnel ID value represents the intended recipient of the packet.

Session ID field
The Session ID field is 16 bits in length and follows the Tunnel ID field. The purpose of the Session ID field is to provide an identifier for a session within a tunnel. Similar to the Tunnel ID field, the Session ID field value has local significance and the same session will be assigned different Session ID values by each endpoint.

Ns and Nr fields
The Ns and Nr fields are both optional, with their presence indicated by the setting of the S bit. Ns indicates the sent sequence number for a data or control message, while Nr indicates the sequence number expected to be received. Thus, Nr is set to the value of the last received message Ns field value plus 1.

Offset Size field

The Offset Size field is also optional, with its presence indicated by the setting of the 0 bit. When present, the value of this field indicates the number of bytes past the L2TP header at which the actual payload commences.

4.4.4 Control messages

The setting of the Type (T) bit indicates that the L2TP message is a control message. Similar to PPTP, L2TP control messages fall into distinct message categories. Those categories include connection management, call management, error reporting and PPP session control. As we briefly turn our attention to the four types of L2TP control messages, we need to digress a bit and discuss the manner by which messages are encoded, a technique referred to as Attribute Value Pair or AVP.

4.4.4.1 Attribute Value Pairs

Under L2TP, Attribute Value Pairs represent a uniform mechanism for encoding different message types. The format of an AVP is shown in Figure 4.26. The AVP represents the variable length concatenation of a unique Attribute that is represented by an integer and an Attribute Value that contains the actual value identified by the attribute. Multiple AVPs are used to develop control messages used for creating, maintaining and tearing down L2TP tunnels.

M-bit field

The setting of this 1-bit field controls the behavior of an implementation that receives an unrecognized AVP. If the M-bit is set on an unrecognized AVP within a message associated with a particular session, the session associated with the message must be terminated. In contrast, if the M-bit is not set, the unrecognized AVP is ignored and the processing of the control message continues.

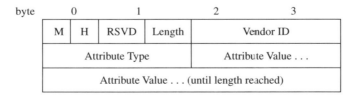

Figure 4.26   Attribute Value Pair (AVP) format.

H-bit field

The Hidden (H-bit) field provides a mechanism to avoid the passing of sensitive data, such as passwords, as cleartext in an AVP. When set, it identifies the hiding of data in the Attribute Value field of the AVP. When this situation occurs several operations occur to obscure the length of the attribute value being hidden, after which an MD5 hash is performed using the shared secret, the 2-byte Attribute number of the AVP and a random vector. The MD5 hash value is then manipulated based upon a hiding method adapted from RFC 2138.

AVP length field

This 10-bit field indicates the number of bytes in the AVP. The length is calculated as 6 plus the length of the Attribute Value field and can vary from 6 to 1023, with a value of 6 indicating the AVP Value field is absent.

Vendor ID field

This 16-bit field is set to a value of 0 to indicate IETF adopted attribute values. Vendors wishing to implement L2TP extensions would use their own Vendor ID along with private attribute values to prevent a collision with another vendor's extensions.

Attribute Type field

This 16-bit field identifies the AVP for a given vendor ID value. Table 4.8 lists presently defined IETF adopted (AVP Vendor ID value of 0) AVP Types.

TABLE 4.8   IETF defined AVP Types

Type	Description
1	Message Type
2	Result Code
3	Protocol Version
4	Framing Capabilities
5	Bearer Capabilities
6	Firmward Revision
7	Host Name
8	Vendor Name
9	Assigned Tunnel ID
10	Receive Window Size
11	Challenge

(*continued overleaf*)

TABLE 4.8 (*continued*)

Type	Description
12	Q.931 Cause Code
13	Response
14	Assigned Session ID
15	Call Serial Number
16	Minimum BPS
17	Maximum BPS
18	Bearer Type
19	Framing Type
20	TBD
21	Called Number
22	Calling Number
23	Sub-Address
24	Tx Connect Speed BPS
25	Physical Channel ID
26	Initial Received LCP CONFREQ
27	Last Sent LCP CONFREQ
28	Last Received LCP CONFREQ
29	Proxy Authen Type
30	Proxy Authen Name
31	Proxy Authen Challenge
32	Proxy Authen ID
33	Proxy Authen Response
34	Call Errors
35	ACCM
36	Random Vector
37	Private Group ID
38	Rx Connect Speed
39	Sequencing Required
40	Rx Minimum
41	Rx Maximum
42	Service Category
43	Service Name
44	Calling Sub-Address
45	VPI/VCI Identifier
46	PPP Disconnect Cause Code
47	CCDS
48	SDS
49	LCP Want Options
50	LCP Allow Options
51	LNS Last Sent LCP Confreq
52	LNS Last Received LCP Confreq

Attribute Value field
This field contains the actual value indicated by the Vendor ID and Attribute Type fields. For example, from Table 4.8 a Vendor ID of 0 and Attribute Type value of 1 would result in the Attribute Value field containing a result code, indicating the reason for terminating the control channel or session. This variable length field has a length equal to the value in the Length field less 6 bytes of the header and is absent when the Length field has a value of 6.

If you carefully examine RFC 2661 you can note how some AVPs are closely related to both PPTP and L2F. For example, the Result Code AVP can contain optional Error Code and Error Message fields. The Error Code field is a 16-bit integer that defines a predefined coded reason for a problem, while the Error Message contains a string that provides text associated with the condition. Similar to L2F, L2TP has a series of general error codes as well as a specific code value that requires additional information about the error to be included in the Error Message field.

4.4.4.2 L2TP Control Message Types

Similar to PPTP and L2F, L2TP supports a series of control messages that can be grouped into different categories. Table 4.9 lists presently defined L2TP control messages.

The actual operation of a control message is specified through the use of different AVPs. For example, an AVP Attribute Type 2, which indicates the L2TP protocol version (see Attribute Type 2 in Table 4.8), is used in both SCCRQ and SCCRP control messages. Similarly, the framing capabilities (Attribute Type 3) is used to define the type of framing (asynchronous, synchronous or both) supported. Rather than repeat detailed information from the RFC that defines the structure and use of each AVP and the control messages they can be used with, they are summarized in a unique and hopefully more valuable table for reference than afforded by the RFC. Table 4.10 summarizes the L2TP AVPs by their applicability to different control messages.

4.4.5 Protocol operations

There are two basic steps associated with tunneling a PPP session under L2TP. First, a control connection for the tunnel needs to be established. Second, a session is triggered by an incoming or outgoing call request.

The establishment of a control connection between an LAC and LNS requires several operations. Those operations include a three-way message exchange or handshake as well as identifying the peer, its L2TP version in use, exchanging

TABLE 4.9 L2TP control message types

Control Connection Management

0	(reserved)	
1	(SCCRQ)	Start-Control-Connection-Request
2	(SCCRP)	Start-Control-Connection-Reply
3	(SCCCN)	Start-Control-Connection-Connected
4	(StopCCN)	Stop-Control-Connection-Notification
5	(Reserved)	
6	(HELLO)	Hello

Call Management

7	(OCRQ)	Outgoing-Call-Request
8	(OCRP)	Outgoing-Call-Reply
9	(OCCN)	Outgoing-Call-Connected
10	(ICRQ)	Incoming-Call-Request
11	(ICRP)	Incoming-Call-Reply
12	(ICCN)	Incoming-Call-Connected
13	(Reserved)	
14	(CDN)	Call-Disconnect-Notify

Error Reporting

15	(WEN)	Wan-Error-Notify

PPP Session Control

16	(SLI)	Set-Link-Info

framing and bearer capabilities and other information. The basic three message handshake is shown in Figure 4.27. If the LAC or LNS requires authentication of its peer, it can include a Challenge AVP in the SCCRQ or SCCRP message. If a Challenge AVP is received, this requires a Challenge Response AVP to be included in the following SCCRP or SCCCN. For both LAC and LNS to participate in tunnel authentication they must be configured with a common single shared secret.

Once a tunnel is established, another troika of messages must be exchanged to establish a session. If it is an incoming call the LAC will transmit an ICRQ (see Table 4.9) message to the LNS. The LNS will respond with an ICRP message and the LAC will return an ICCN message. In contrast, if an outgoing call is being used to set up a session, the LNS will transmit an OCRQ message to the LAC. The LAC will return an OCRP message, perform the call and then generate an OCCN message to indicate the call is connected. L2TP has a comprehensive series of control messages used for sending keep-alives (Hello messages) through session and control connection tear-down. Different AVPs

TABLE 4.10 Control Connection Management AVPs

AVP	Message Applicability		
	SCCRP	SCCRQ	StopCNN
Protocol Version (Type 2)	X	X	
Framing Capabilities (Type 3)	X	X	
Bearer Capabilities (Type 4)	X	X	
Tie Breaker (Type 5)		X	
Firmware Revision (Type 6)	X	X	
Host Name (Type 7)	X	X	
Vendor Name (Type 8)	X	X	
Assigned Tunnel ID (Type 9)	X	X	X
Receive Window Size (Type 10)	X	X	
Challenge (Type 11)	X	X	
Challenge Response (Type 12)	X	X	

AVP	CDN	ICRP	ICRQ	OCRP	OCRQ	ICCN	OCCN
Q.931 (Type 12)	X						
Assigned Session ID (Type 14)	X	X	X	X	X		
Call Serial Number (Type 15)			X		X		
Minimum BPS (Type 16)					X		
Maximum BPS (Type 17)					X		
Bearer Type (Type 18)					X		
Framing Type (Type 19)					X		
Called Number (Type 21)			X		X	X	X
Calling Number (Type 22)			X		X		
Sub-Address (Type 23)			X	X			
Tx Connect Speed (Type 24)						X	X
Rx Connect Speed (Type 38)						X	X
Physical Channel (Type 25)				X	X		
Private Group ID (Type 37)						X	
Sequencing Required (Type 39)						X	X

Proxy LCP and Authentication AVPs	ICCN
Initial Received LCP CONFREQ (Type 26)	X
Last Sent LCP CONFREQ (Type 27)	X
Last Received LCP CONFREQ (Type 28)	X
Proxy Authen Type (Type 29)	X
Proxy Authen Name (Type 30)	X
Proxy Authen Challenge (Type 31)	X
Proxy Authen ID (Type 32)	X
Proxy Authen Response (Type 33)	X

Call Status AVPs	WEN	SLI
Call Errors (Type 34)	X	
ACCM (Type 35)		X

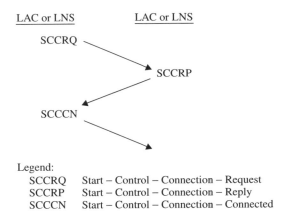

Legend:
SCCRQ   Start – Control – Connection – Request
SCCRP   Start – Control – Connection – Reply
SCCCN   Start – Control – Connection – Connected

Figure 4.27   Establishing an L2TP control connection.

can be present in different messages that make the protocol very complex and require a 102-page RFC to define. Thus, this author will leave it to the reader to probe deeper into the RFC for minute details concerning the role of AVPs in different messages. Instead of further discussing AVPs we will conclude our examination of L2TP by turning our attention to the manner by which the two architectural models of the protocol operate and how data flows under each model.

Earlier in this section we noted that L2TP supports two general architectural models. In concluding this section we will examine in detail the manner by which each model operates to include the flow of data from source to destination.

4.4.5.1 LAC-LNS Model

Under the LAC–LNS model previously illustrated in Figure 4.24, the remote client initiates a connection to their ISP. The ISP accepts the connection, resulting in the establishment of a PPP connection. The ISP will perform a partial authentication to determine the user name of the remote client. The user name is then matched against an ISP database to determine those services available for the user. Assuming the user is predefined for L2TP VPN services, the LAC will initiate an L2TP tunnel to the LNS. If the LNS accepts the connection, the LAC then encapsulates the remote user's PPP data under L2TP and forwards it to the LNS. At the LNS the packets are

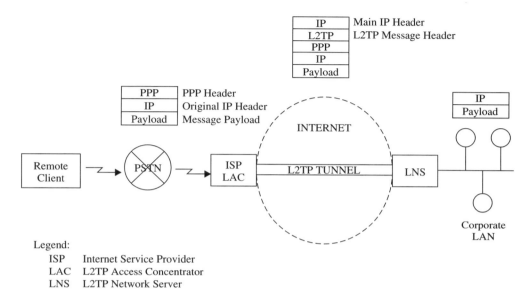

Figure 4.28   L2TP LAC–LNS data flow.

stripped of the L2TP header and processed as PPP data. The LNS then uses PPP authentication to validate the user.

Data flow

Under the previously described model data flow occurs from the remote client to the ISP via PPP. From the ISP, data flow occurs via L2TP to the LNS, after which data flow will occur via IP on the corporate LAN. Figure 4.28 illustrates the flow of data under the L2TP LAC–LNS architectural model.

4.4.5.2 Direct Client-LNS Model

The second architectural model supported under L2TP bypasses the LAC through the use of L2TP client software. Under this architectural model, which was previously shown in Figure 4.24, the remote user has a pre-established connection to an ISP. That connection can be via cable or DSL modem or through the client residing on an organizational LAN connected to the Internet. The remote user operating L2TP client software in effect becomes an LAC and initiates an L2TP tunnel to the LNS. If the LNS accepts

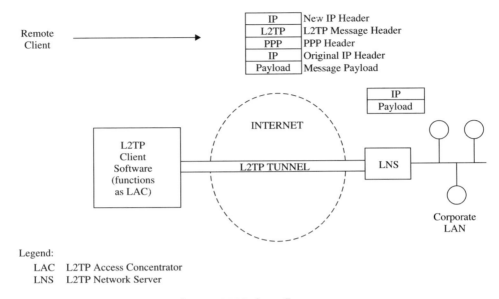

Figure 4.29 L2TP Direct client–LNS data flow.

the connection, the remote client encapsulates PPP within L2TP, forwarding it through the tunnel. At the opposite endpoint the LNS removes the L2TP header and processes the resulting data, using PPP authentication to validate the remote user.

Figure 4.29 illustrates the flow of data under the L2TP direct client to server architectural model. Note that the remote user operating L2TP client software in effect functions as a LAC, although it bypasses any ISP LACs and establishes a tunnel directly to the LNS on the corporate LAN.

chapter five

Higher Layer VPNs

As the title implies, we will now turn our attention to higher layer VPNs. In doing so we will not cover all types of higher layer VPNs, but instead will discuss two practical and popular types of VPNs: IPSec and the Secure Sockets Layer (SSL). First, we will examine the IPSec protocol, which more accurately represents two protocols and a mechanism for key exchange. Then we will turn our attention to the use of IPSec with L2TP, since the latter by itself does not provide a built-in mechanism to secure the tunnel created through its use.

In the second section in this chapter we will examine the configuration of a server to support IPSec. In doing so we will use a Windows 2000 server, since that platform represents the most popular server platform in use when this book was written. In concluding this chapter, we will turn our attention to SSL, examining the flow of messages used to perform authentication and encryption and the use of appliance hardware, which enables a browser client without modification to be used to create a secure connection to a variety of TCP/IP applications.

5.1 Understanding IPSec

IPSec represents a set of protocols that provide cryptographic security, authentication, data integrity, access control and confidentiality. Under IPSec you can transmit data that secures applications from eavesdropping as well as from modification. Although the use of IPSec is by design transparent to communications, this set of extensions to the IP protocol family must operate on both ends of a communications link and, as we will note later in this chapter, is primarily used to secure the creation of a VPN through the Internet by its use with a tunneling protocol.

Virtual Private Networking   G. H. Held
© 2004 John Wiley & Sons, Ltd   ISBN: 0-470-85432-4

5.1.1 Overview

IPSec dates back to 1992 when the IP Security Protocol Working Group (IPSec) was formed by the Internet Engineering Task Force (IETF). This working group was tasked with developing a standardized method for implementing privacy and authentication services for IPv4 and the then emerging Version 6 (IPv6) of the IP protocol.

The working group assigned to the development of IPSec produced a flexible security architecture that can be used to support both IPv4 and IPv6 as well as all IP protocols, such as TCP, UDP, and even ICMP. In addition, the flexibility of IPSec enables it to be used to secure communications between a pair of hosts or multiple hosts, two networks or multiple networks or a combination of hosts and networks. Here we are using the term 'networks' to denote a large number of hosts residing on a specified network or subnet. As we will note later in this section, when secure communications to a network occur, IPSec flows to a gateway device that functions as a proxy for the hosts on the network or subnet behind the gateway. That gateway can be a stand-alone device or a router or firewall that supports IPSec.

Because the securing of multiple hosts or networks can result in a major effort for the distribution of encryption keys, another goal of IPSec was to develop a method for automatically distributing such keys. The resulting work of the IPSec working group is a series of RFCs, mainly from 2401 to 2412 that were published in 1998.

5.1.2 Topologies supported

The need to secure communications between hosts and networks or a combination of the two resulted in a requirement for IPSec to support two connection topologies. Those topologies are host-to-host and gateway-to-gateway. The host-to-host topology provides a direct end-to-end secure communications path between two devices. In comparison, the gateway-to-gateway model, which is also referred to as a network-to-network model, results in two gateways functioning as security proxies on behalf of trusted hosts connected to each gateway's internal, trusted network.

The gateways establish one or more tunnels between themselves and other IPSec gateways. Then, the destination address in the IP header of each datagram generated by a host flowing onto another network is examined by the gateway serving the host. If the destination address resides on a network associated with a distant organizational gateway that supports IPSec, the datagram will be manipulated to flow in a secure manner to that gateway. In comparison, if the destination network address in the datagram differs

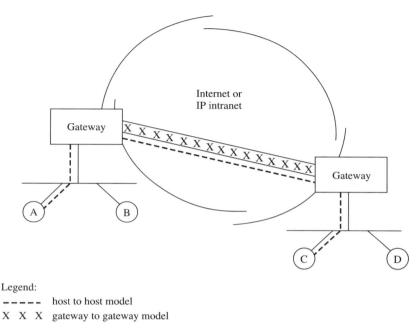

Figure 5.1   Basic IPSec topologies.

from that of a remote gateway the local gateway allows the datagram to pass unaltered.

Figure 5.1 provides a comparison of host-to-host and gateway-to-gateway IPSec topologies. Under the host-to-host model, security is truly end-to-end, whereas under the gateway-to-gateway model communications from the host to its gateway are not secure. Recognizing this problem resulted in several variations to the basic topology shown in Figure 5.1.

One common variation enables hosts behind a gateway to transmit data securely to the gateway. Another variation permits remote hosts to directly communicate with a gateway. As we look closer at IPSec we will also examine its basic topologies and variations of those topologies.

5.1.3 Specifying session parameters

IPSec is similar to other security protocols in that a mechanism is required to specify session parameters. Under IPSec the method used to specify session parameters is divorced from any specific algorithm, providing a flexible mechanism for not only specifying session parameters but also providing a structure that can be used to support new algorithms as they are developed. The mechanism used by IPSec is referred to as a Security Association (SA), which can be viewed as an undirectional, logical connection between two IPSec systems.

In addition to functioning as a logical connection between two IPSec systems, the SA has physical attributes. Those physical attributes are in the form of a table or database record of security parameters that govern operations between two IPSec compliant systems. SA tables are established on a destination host and referenced by a sending host through the use of an index parameter. That parameter is referred to as a Security Parameter Index (SPI). Thus, an SA can be identified by the following triplet:

<Security Parameter Index, IP Destination Address, IP Source Address>

5.1.3.1 Security Association Entries

As briefly mentioned earlier in this section, the SA table consists of a set of security parameters that govern the manner by which data is protected. Examples of SA entries in a table or database record can include the following items:

- The type of cryptologic transformation to be used, such as DES, Triple-DES, or the recently standardized AES.
- The key or keys used by the cryptographic transformation. Such keys could be entered manually when the SA is defined on a host or gateway. Alternately, keys could be provided via a distribution system or if a public key system is used, transmitted during the connection setup process.
- Any initialization vector or encryption synchronization required to initialize the cryptographic process.
- The duration of the cryptographic keys.
- The duration of the security association.
- The source address of the security association.

Concerning the above examples of SA entries, it is important to note that there is no predefined life span for cryptographic keys. Thus, the frequency at which keys are changed are at the discretion of the person configuring an endpoint. However, because it is easier to break a system that uses a key for a long period of time than a system where keys are periodically changed, it is a good idea to periodically change keys. Similarly, although there is no predefined life span for a security association you may wish to periodically change the SA.

5.1.4 The SPI

The Security Parameter Index (SPI) is a 32-bit pseudo-random number included in the header of each of the two security protocols supported by IPSec. Those protocols, referred to as Authentication Header (AH) and Encapsulating Security Payload (ESP) will be covered in detail later in this section.

An SPI value can be entered manually when an SA is defined on a host or gateway or can be provided via an SA distribution system. To work correctly, SPIs must be synchronized between endpoints as their use references a table of security parameters that must work in tandem at each endpoint. SPI values in the range 1 to 255 are reserved by the Internet Assigned Numbers Authority (IANA) for use with standard implementations, while a value of 0 indicates the absence of a Security Association for a given transaction.

The actual use of the SPI depends upon the architectural model of IPSec. In a host-to-host connection the SPI is used by the receiving host to directly look up the Security Association. In a gateway-to-gateway topology the SPI is combined with the destination address to determine the applicable SA since the gateway supports multiple endpoints.

5.1.5 Protocols

The ability of IPSec to provide confidentiality, data integrity, and authentication results from the use of two security related protocols referred to as Authentication Header (AH) and Encapsulated Security Payload (ESP). The use of AH provides both integrity and authentication for IP datagrams, however, it does not support data integrity through encryption. In contrast, ESP can be used to provide integrity, authentication and encryption to IP datagrams. A third protocol, referred to as the Internet Key Exchange (IKE) provides a mechanism to negotiate the use of encryption keys. Unlike AH and ESP, IKE is optional since it is possible to manually configure encryption keys. However, from a protocol standpoint it is doubtful if you would want to periodically change keys for a multi-site IPSec system and in this scenario you would more than likely use IKE.

5.1.5.1 Modes of Operation

Both AH and ESP support two modes of operation referred to as transport mode and tunnel mode. In the transport mode, the applicable protocol header (AH or ESP) is inserted directly behind the IP header as illustrated in the top portion of Figure 5.2. In the tunnel model the applicable protocol header (AH or ESP) is prefixed to the original IP datagram, resulting in the original

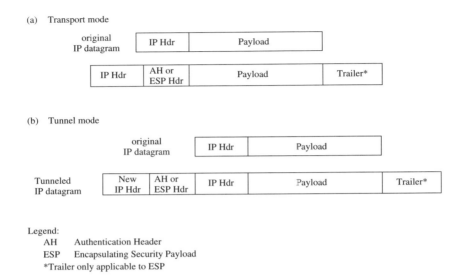

(a)   Transport mode

(b)   Tunnel mode

Legend:
AH   Authentication Header
ESP   Encapsulating Security Payload
*Trailer only applicable to ESP

Figure 5.2   Transport and tunnel mode packet manipulation.

datagram actually becoming a payload. A new IP header is then prefixed to the IPSec header to create a tunneled payload. The general format of a tunneled IP datagram is shown in the lower portion of Figure 5.2.

In comparing transport and tunnel modes, their advantages, disadvantages and utilization are similar regardless of the IPSec protocol used. The transport mode was defined to be used by hosts, while the tunnel mode is used whenever either end of a security association is a gateway. Although a gateway is supposed to be limited to tunnel mode and is not required to support transport mode, many do so by necessity. For example, when a gateway supports transport mode it functions as a host, enabling the gateway to respond to IPSec packets addressed to itself. Doing so enables the gateway to respond to SNMP commands as well as various types of ICMP messages. A key difference between transport mode and tunnel mode is the prefix of a new IP header onto a datagram when the tunnel mode is employed. While this action requires more processing overhead, it enables datagrams transmitted by different hosts to be directed to a common gateway.

Now that we have a general appreciation of the manner by which transport mode and tunnel mode operations modify AH and ESP headers, let us turn our attention to those headers. In doing so we will note the fields within these headers and their use as well as how the headers are actually used in transport mode and tunnel mode.

5.1.6 Authentication Header

The Authentication Header (AH) is used to provide data integrity and authentication to IP datagrams. To do this, IPSec authentication occurs by default using Message Digest Version 5 (MD5). MD5's hash or another algorithm specified in the security association is computed over the entire datagram. Because the algorithm is applied to a datagram prior to any required fragmentation occurring, when fragmentation does occur, certain fields, such as the flags and checksum, will change. In addition, other fields in the IP header, such as the Time-To-Live (TTL) or hop count and Type of Service (TOS) can also change and are not predictable by the receiver. Due to this, such fields are excluded from the computation and are not protected by the Authentication Header. Such fields are referred to as mutable and are listed in Table 5.1.

Once the hash has been computed, it is inserted into the Authentication Header (AH) along with the SPI assigned to the security association. The top portion of Figure 5.3 illustrates the placement of the Authentication Header within an IP datagram. Note that the Authentication Header is inserted into the datagram right after the IPv4 header or the hop-by-hop header under IPv6 and is identified by the IP header having a value of 51 in its Protocol field. By placing authentication data in its own inserted header, the original datagram is unaltered. This means it can be read and processed by systems that are not participating in the authentication process. The lower portion of Figure 5.3 shows the fields of the AH header. As indicated by the illustration the AH header consists of six fields.

Next Header field
This 8-bit field identifies the type of the next payload after the AH header. The Internet Assigned Numbers Authority (IANA) defines the assigned numbers that can be used.

Payload length field
This 8-bit field contains the length of the AH header expressed in 32-bit words, minus two.

TABLE 5.1   Mutable IPv4 fields

Type of Service (TOS)
Flags
Fragment Offset
Time-to-Live (TTL) or hop count
Header Checksum

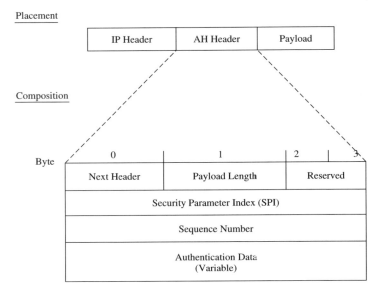

Placement

Composition

Byte

Figure 5.3   The IPSec authentication header.

Reserved field

This 16-bit field is reserved for future use. It is currently set to zero.

Security Parameter Index field

This 32-bit field identifies the security association to be used for processing the datagram.

Sequence number field

This 32-bit field is initially set to zero when a security association is established. Thereafter, before each datagram is transmitted, the value of this field is incremented by 1. Because sequence numbers associated with a security association are not allowed to repeat when the sequence number value of $2^{32} - 1$ is reached, a new SA must be established.

The use of the sequence number field provides protection against a third party monitoring and replaying data. This protection, referred to as replay protection, is optional although the field is always included by the sender. It is up to the receiver to determine whether or not to process the data in this field.

Authentication Data field

This variable length field holds the data output by the cryptographic algorithm. That algorithm is selected for use during the security association initialization process. Currently most implementations of IPSec support at least two cryptographic algorithms, such as MD5 and SHA-1.

A third authentication algorithm is the Hashed Message Authentication Code (HMAC), which is used in conjunction with MD5 to enhance the strength of the MD5 algorithm. This algorithm is commonly referred to as MMAC-MD5.

At the receiver the same algorithm is applied to the data to verify its integrity. Thus, another name for this field is the Integrity Check Value (ICV). If the receiver's computation matches the contents of the Authentication Data field, the datagram is presumed to have arrived both without error and tampering.

5.1.6.1 AH Modes

As noted earlier during our overview of IPSec, both AH and ESP support two modes of operation: transport mode and tunnel mode. In this section we will focus our attention upon the use of those modes by the Authentication Header protocol.

Figure 5.4 provides a comparison of the transport of an IP datagram under AH operating in transport mode and tunnel mode. Under the transport mode, which is used by hosts, the AH header is inserted after the IP header and authenticates the newly formed datagram less those fields in the IP header whose values can change and which are referred to as mutable fields. As a review, in the tunnel mode of operation, at least one end of a security

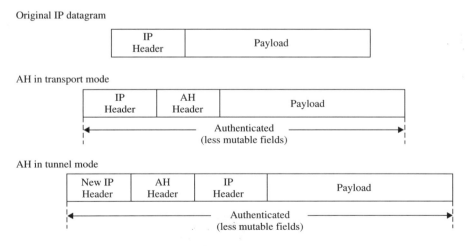

Figure 5.4 Comparing AH modes of operation.

association is a gateway. This requires the original datagram to be tunneled or encapsulated within a new IP header that routes the original datagram to the gateway. When the Authentication Header protocol is used in the tunnel mode of operation, its header is inserted between the new IP header and the original IP header as shown in the lower portion of Figure 5.4.

Note that because routing occurs based upon the new IP header the contents of the original header do not change. This means there are no mutable fields in the original IP header and all fields are then authenticated in that header. A second advantage associated with AH in tunnel mode as well as ESP in tunnel mode concerns the use of IP addresses. Because the new IP header is used for routing, the original header can contain private RFC1918 addresses. This means that the use of tunnel mode enables you to secure transmission to or from a network with private IP addresses without having to worry about address translation. However, because you cannot route data between two networks having the same IP address, you cannot configure the same RFC 1918 private network address on each network to be interconnected via IPSec tunnel mode.

Now that we have an appreciation of AH and its modes of operation, let us turn our attention to ESP.

5.1.7 Encapsulating Security Payload

The Encapsulating Security Payload (ESP) adds the ability to encrypt IP datagrams to the integrity checking and authentication capability provided by AH. Thus, ESP protects your data from viewing as well as insures it arrived without error and tampering. Because ESP adds encryption, it appears that the designers of this protocol decided to enable authentication to occur without decryption. However, to accomplish this required the authentication data to be placed outside the encrypted data in the form of a trailer, resulting in a more complicated packet format than AH. As we will shortly note, ESP requires a header and two trailers, with the payload encapsulated between the header and trailers which contributes to the name of the protocol.

Figure 5.5 illustrates the fields within the ESP Header and Trailers. The top portion of Figure 5.5 indicates the general structure of an IP datagram in transport mode, showing the relationship of the header and trailers to the datagram.

Similar to AH, ESP is applicable to both IPv4 and IPv6 and is identified by a specific value in the Protocol field of the IP header. That value is 50, while a value of 51 in the Protocol field identifies AH.

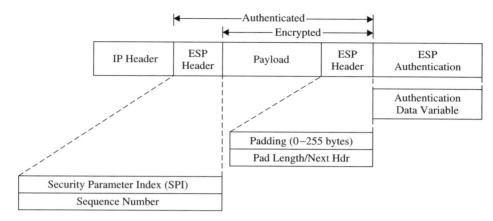

Figure 5.5   ESP header and trailer fields.

The lower portion of Figure 5.5 indicates the fields within the header and trailers. We can obtain an appreciation of ESP by examining the fields in its header and trailers, so let us do so.

Security Parameter Index field
The Security Parameter Index (SPI) field is 32-bits in length. Similar to the SPI in an AH header, this field is used to identify different Security Associations (SAs) with the same destination address and security protocol.

Sequence number field
This 32-bit field functions in the same manner as its counterpart in the AH header. That is, it increments by 1 per packet as a mechanism to provide protection against a replay. After the highest sequence number ($2^{32} - 1$) is used, a new Security Association must be established as sequence numbers for a given SA cannot repeat.

Payload field
This variable length field contains the actual data being transported. This field, as well as any padding and the Pad Length and Next Header fields to be described shortly, are encrypted, with the algorithm used for encryption selected when the Security Association was established. The actual type of data carried in the payload, such as a TCP segment, is identified by the Next Header field contained in the trailer.

5.1.7.1 ESP Trailer

As indicated in Figure 5.5, there are three fields in the ESP trailer. Those fields include an optional Padding field as well as mandatory Pad Length and Next Header fields.

Padding field

The presence of this optional variable length field becomes necessary when the Next Header field is otherwise not aligned to terminate on a 4-byte boundary. In addition, this field permits the use of encryption algorithms that operate on a block basis, permitting the variable payload to be extended to a block boundary.

Pad Length field

The purpose of this 8-bit field is to define the number of padding bytes preceding this field. Although the padding field can be omitted when the payload data falls on a cryptographic byte boundary, the Pad Length field is mandatory. In this situation the value of the Pad Length field would be set to 0 to indicate no padding.

Next Header field

The purpose of this 8-bit field is to indicate the type of data carried in the Payload field. The values for this field correspond to the set of IP protocol numbers defined by the IANA. For example, a value of 6 would indicate a TCP segment while a value of 16 would indicate a UDP datagram.

5.1.7.2 ESP Authentication

If we return to Figure 5.5 we can note a second ESP trailer. That trailer is ESP Authentication, which consists of a single field of Authentication Data.

Authentication Data field

The Authentication Data field is variable in length but ends on a 4-byte boundary. This field contains the ICV computed from the SPI field at the beginning of the ESP header through the Next Header field in the trailer. Thus, authentication under ESP covers the ESP Header, Payload and ESP Trailer but does not cover the IP header. It should be noted that this field is only present when integrity check and authentication are selected when the Security Association is established. Otherwise this field will not be included in the ESP packet.

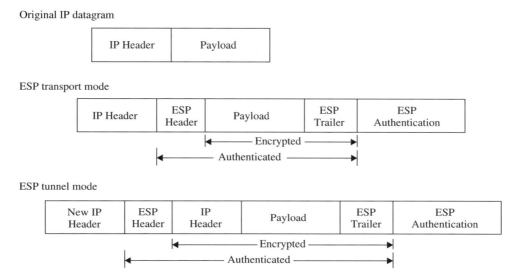

Figure 5.6   ESP modes of operation.

5.1.7.3 ESP Modes

Similar to AH, ESP can operate in two modes: transport mode and tunnel mode. Figure 5.6 provides a comparison of an original IP datagram carried under ESP transport mode and ESP tunnel mode. From a visual comparison of the two modes it is obvious that the transport mode provides neither authentication nor encryption for the IP header. However, it is important to remember that the transport mode is used by hosts communicating with one another. Therefore, it would be impossible to encrypt the header since routers along the path between endpoints need to operate upon certain fields in the header. In comparison, under the tunnel mode of operation the original IP header is both encrypted and authenticated, providing total protection for the original IP datagram.

This protection is possible because a new header is prefixed to the datagram, which enables the contents of the original header to be encrypted. Similar to AH, ESP in tunnel mode is used when a gateway is at either end of a security association. While gateways under IPSec are assumed to only operate in tunnel mode, under certain situations they may support a transport mode of operation. For example, when a gateway must operate upon traffic directed to its IP address, such as responding to SNMP commands, it can be expected to operate in a transport mode.

5.1.8 Operations

Earlier when we discussed topology we noted that IPSec supports two basic topologies: host-to-host and gateway-to-gateway. We also briefly noted that it is possible for a host to communicate directly with a gateway. Since a gateway provides access to a network, from a logical perspective IPSec can work in three different ways. Those ways are: Host-to-Host; Host-to-Network; and Network-to-Network.

5.1.8.1 Host-to-Host

Host-to-host operations result in an IPSec connection between two hosts and represent an end-to-end topology. Either transport mode or tunnel mode can be used under host-to-host operations. In the transport mode you can use AH or ESP, or ESP can be applied after AH, a situation referred to as transport adjacency, which we will shortly examine. Both AH and ESP by themselves can be supported under tunnel mode. Figure 5.7 illustrates an example of a host-to-host operation, where a host on one network directly communicates with a host on another network. Note that the connection between the two hosts is considered to represent an IPSec tunnel even when transport mode is used. Thus, it is important to distinguish the physical mode of operation (transport or tunnel) from the logical tunnel formed from the use of either mode of IPSec communications.

5.1.8.2 Host-to-Network

In a host-to-network environment a gateway provides IPSec support for multiple clients that can be on the same network or located on different

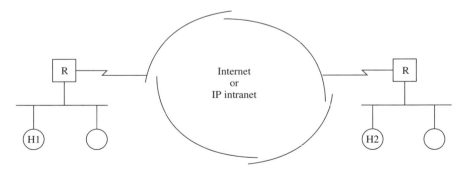

Connection between H1 and H2 represents an IPSec tunnel.

Figure 5.7   Host-to-Host operation.

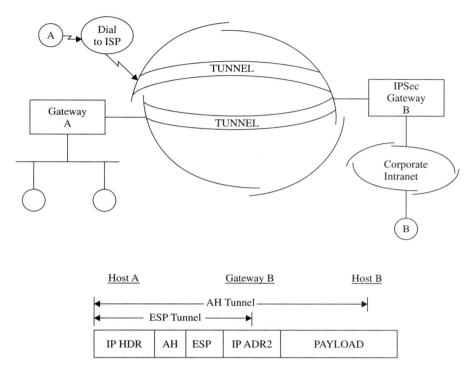

Figure 5.8   Host-to-Network operation.

networks. Each client host communicates with the gateway instead of a particular host on a destination network. Because one endpoint is a gateway, only tunnel mode can be supported, however, that mode can occur under either AH or ESP.

Figure 5.8 illustrates an example of a host-to-network transmission model. In this example each remote host would use AH or ESP in tunnel mode to the gateway. It is also possible to create a combined AH-ESP tunnel whose data flow and packet header structure is shown in the lower portion of Figure 5.8.

Note that from host A to gateway B, AH and ESP are shown combined. Since an ESP header cannot authenticate an outer IP header, it is useful to combine both AH and ESP as shown in the lower portion of Figure 5.8. Although this combination appears to represent a version of transport adjacency, from a technical perspective it is actually referred to as iterated tunneling. Later in this section we will turn our attention to both a formal definition of transport adjacency as well as an examination of the difference between transport adjacency and iterated tunneling.

5.1.8.3 Network-to-Network

In a network-to-network operational environment hosts on one network communicate with a gateway that operates the IPSec protocol stack. That gateway communicates with a second gateway connected to the destination network that also operates the IPSec protocol stack. Clients on each network do not need to have any knowledge of IPSec as the gateways automatically perform all security-required operations.

Figure 5.9 illustrates an example of the establishment of a VPN tunnel from one organizational intranet to another. In this example the two gateways, which can represent routers with IPSec capability or separate devices support tunnel mode, with either AH or ESP used. While AH requires less processing overhead on the gateway, its use does not hide the contents of data from observation as it flows across the Internet. Thus, most organizations that require data confidentiality would prefer the use of ESP between gateways.

5.1.8.4 Variations

One of the problems associated with network-to-network operations is the fact that data is only authenticated or authenticated and encrypted between gateways. While network-to-network operations eliminate the necessity to support IPSec on each host, security only occurs from gateway-to-gateway. If

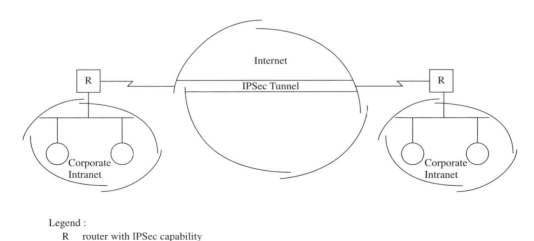

Legend :
R   router with IPSec capability
◯   workstation

Figure 5.9 Using gateways at each network provide the capability for a network-to-network IPSec capability.

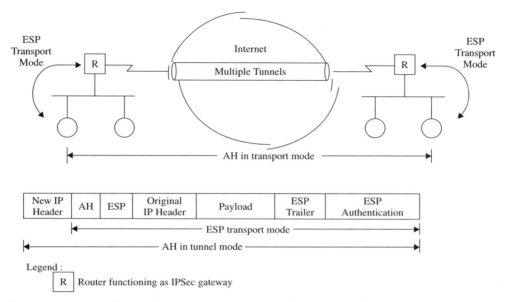

New IP Header	AH	ESP	Original IP Header	Payload	ESP Trailer	ESP Authentication

Figure 5.10   End-to-end security via network-to-network operations.

an organization requires end-to-end security, you should consider a variation of network-to-network operations if your organization requires a literal 'last mile' level of protection. For example, you could configure hosts that require 'last mile' protection to use both AH and ESP in transport mode to the distant host. This action would both authenticate and encrypt IP datagrams as they flow to the local gateway for routing to their destination. Figure 5.10 illustrates the operation of end-to-end security via network-to-network operations.

In examining Figure 5.10 note that hosts requiring end-to-end security use AH and ESP in transport mode, resulting in authenticated and encrypted packets flowing through the local gateway as a combined AH-ESP tunnel to the remote host. This tunnel, whose datagram composition is shown in the lower portion of Figure 5.10 results in a new IP header and an AH header prefixing the ESP transport mode datagram. The fact that two protocols are used means you must have two Security Associations, since a Security Association can only involve one security protocol – AH or ESP. Thus, in concluding our discussion of networking variations let us turn our attention to transport adjacency, security associations and iterated tunneling to obtain an understanding of the relationships between this troika of IPSec terms.

5.1.8.5 SA Bundles

Earlier we noted that IPSec connections are defined in terms of Security Associations (SA). As a quick review, a Security Association is defined for a single unidirectional flow of packets from one point to another based upon defining three fields: Destination IP address, Security Parameter Index (SPI) and Security Protocol. Because different sources transmitting to a common destination can require different SPIs or security protocols, SAs can be expected to vary. However, if two gateways communicate using a single SA, all traffic flowing between the gateways can then be treated in the same manner. When we extend IPSec on an end-to-end basis, it is possible for some traffic to be subject to several SAs, each of which applies some transformation to the datagram by the insertion of a header, trailer and for ESP the manipulation of the payload based upon a cryptographic algorithm. In fact, we noted that because a SA can only involve one protocol, the use of AH and ESP would require two SAs when the protocols are combined for use over the entire transmission path. Groups of Security Associations are referred to as an SA Bundle and can be considered to represent a combination of SAs that apply to the same traffic flow.

There are two methods that can be used to create SA Bundles. One method is referred to as transport adjacency, while the other method is known as iterated tunneling. Because they are similar but have subtle differences, let us examine each in detail.

5.1.8.6 Transport Adjacency

The term transport adjacency is used to refer to the application of more than one security protocol to a common IP datagram without using tunnel mode for communications. According to this definition AH and ESP are used to provide a single level of protection without requiring the nesting of communications since the final destination of the datagram is its endpoint. Transport adjacency occurs under host-to-host communications when each host is located behind a gateway. Figure 5.11A illustrates transport adjacency, with the destination responsible for both decryption and authentication.

5.1.8.7 Iterated Tunneling

In comparison to transport adjacency that represents multiple security protocols applied in transport mode, iterated tunneling represents the application of multiple security protocols within a tunnel mode security association, resulting in tunnels within tunnels. Although the differences in the definitions

A . Transport Adjacency.

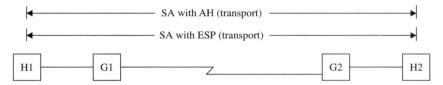

B . Iterated tunneling with identical endpoints.

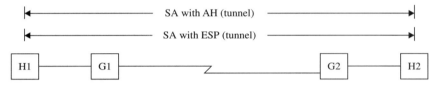

C . Iterated tunneling with one identical endpoint.

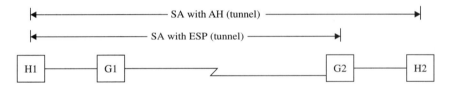

D . Iterated tunneling with no identical endpoint.

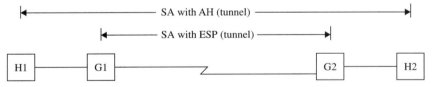

Figure 5.11 Transport adjacency and iterated tunneling.

appear trivial, in reality they are significant. This is because a Security Association can originate or terminate at different points in an end-to-end data flow, enabling multiple layers of nesting or iterated tunneling to occur which allows different endpoints to be associated with each nested tunnel.

Iterated tunneling can occur under three conditions. The first condition is when the endpoints of each SA are identical. A second condition under which iterated tunneling can occur is when one of the endpoints of the SA is the same. For the third condition for iterated tunneling neither endpoint of the SA is the same.

The situation where the endpoints of a Security Association are identical can occur when SAs terminate at the hosts and the hosts are each located behind a gateway as shown in Figure 5.11B. Each of the SAs are end-to-end and the two SAs terminate at the hosts.

When only one of the endpoints is common to the SAs, the security association can be established between the host and remote gateway as well as between the host and another host behind the remote gateway. Figure 5.11C illustrates an example of iterated tunneling with one identical endpoint.

When the situation occurs where neither security association terminates at the same location, it is possible to establish SAs between two gateways and two hosts behind the gateways. An example of this is shown in Figure 5.11D. From a practical standpoint, this configuration could represent a VPN established between two gateways based upon ESP in tunnel mode. Hosts on each network obtain end-to-end secure communications by the use of client-to-client security associations, resulting in an AH tunnel iterated with an ESP tunnel.

Now that we have a better appreciation of the relationship between SA bundles, transport adjacency and iterated tunneling, let us turn our attention to the third pillar of IPSec – key management.

5.1.9 Key management

Until now, our primary examination of IPSec was upon the AH and ESP protocols. The third leg of IPSec is its key management capability. While key management may not be an issue when you use a few gateways or interconnect a handful of hosts securely, for an international or large national network with numerous locations, the management of keys to include their generation, authentication and distribution can represent a considerable effort.

5.1.9.1 Keying Methods

There are two primary methods that can be used to manage IPSec keys: host-oriented and user-oriented. Under a host-oriented key management method, all users share the same key when transmitting data between endpoints, such as two hosts, a host and a gateway or multiple gateways. Under a user-oriented key management method a separate key is used for each user session. This means that a Web session followed by an FTP session would require two keys since each application represents a separate session. Because a user-oriented key management system provides the ability to group users as well as assign different security levels based upon users, user-groups, or applications, this method can be considered to represent a multilevel security (MLS) system. In fact, due to the ability to establish different levels of trust under a user-oriented key management system, the IETF Security Working Group recommends this method for IPSec key management implementations.

5.1.9.2 Relationships

Key management under IPSec represents a combination of protocols referred to as ISAKMP and Oakley that operate under a third protocol referred to as IKE.

ISAKMP

The Internet Security Association and Key Management (ISAKMP) protocol is defined in RFC2408. ISAKMP can be considered to represent a framework that defines the payloads and procedures required to authenticate a peer and enable encryption keys to be generated.

Oakley

Because ISAKMP is a framework that defines payloads for key exchange and authentication, it does not actually define how the key exchange occurs. The protocol used for the key exchange process is referred to as Oakley, which is commonly referred to as ISAKMP/Oakley, as it represents a key exchange protocol that supports the ISAKMP framework. Oakley runs over the UDP protocol using a well-known port.

IKE

The term Internet Key Exchange (IKE) is a protocol defined in RFC2409. IKE can be considered to represent a hybrid protocol that combines portions of ISAKMP and Oakley to provide a key management capability. The terms IKE and ISAKMP are used interchangeably by some vendors while other vendors use the latter term to describe keying functions. While this use is technically correct, it should be noted that ISAKMP provides the framework while Oakley defines the process for the key exchange. Thus, IKE probably represents a better overall term to use when referencing IPSec key management.

5.1.9.3 Message Exchange

The ISAKMP/Oakley key exchange protocol uses a two-phase approach. Under the first phase a set of negotiations occurs that establishes a secure, authenticated communications channel as well as deriving a master secret from which all cryptographic keys are subsequently derived. Commonly, public key cryptography is employed to establish an initial ISAKMP security association between devices as well as to generate keys used to protect messages that flow during the second phase. Thus, you can view Phase One as establishing the foundation for protecting ISAKMP messages, however, no actual security associations or keys are established for protecting user data.

In Phase One there are two modes of operation defined under IKE. Those modes are referred to as main mode and aggressive mode. Each mode is used to establish a Phase One secure exchange. Main mode provides a series of six messages for authentication and is not optional. In comparison, aggressive mode is an option that enables vendors to provide additional information through the use of a minimum message exchange.

In Phase Two communicating devices negotiate the security associations and keys that will be used to protect user data. Phase Two ISAKMP messages are protected by the initial ISAKMP security association generated during Phase One negotiations.

Similar to Phase One, there are two modes of operation defined under Phase Two. Those modes are referred to as quick mode and new group mode. Quick mode is employed to establish security associations. Although new group mode is used to benefit Phase One operations by supporting the creation of additional security associations, this takes place during Phase Two and represents a Phase Two operation.

In comparing phases, it should be noted that Phase One represents an initial or startup exchange that is only periodically performed, perhaps on a daily or weekly basis. In comparison, Phase Two can occur on a session basis and can be expected to occur throughout the day.

5.1.9.4 Establishing a Security Association

Previously we noted that the ISAKMP key exchange represents a two-phase process. The first phase requires the exchange of six messages to establish an initial security association and exchange keys between two IPSec compliant devices. Once Phase One is completed, another series of messages must be exchanged under Phase Two to define the security associations and keys that will be used to protect IP datagrams transmitted between endpoints.

Phase One

Under Phase One the security associations that protect ISAKMP messages are established. As previously mentioned, six messages are required. The

IP Header	UDP Header	ISAKMP Header	Security Association	Proposal 1	Transform Payload for 1	. . .	Proposal N	Transform Payload for N

Figure 5.12   Format of an initial ISAKMP phase one message.

first pair of messages is used to negotiate the characteristics of the security association and provide the foundation for establishing a communication policy. Figure 5.12 illustrates the format of the initial message of an ISAKMP Phase One exchange.

Note that the ISAKMP message is transported as the payload of a UDP segment, which is in turn transported as an IP datagram. The source and destination addresses in the IP header represent the two IPSec endpoints establishing a security association. The UDP header uses the destination port number 500 to identify that it transports an ISAKMP message. Thus, if the two endpoints want to communicate between routers and/or firewalls, you need to permit UDP port 500 destination messages to flow through such devices. Otherwise, your IPSec endpoints will not be able to exchange the messages necessary to establish the security associations to protect ISAKMP messages.

The initial message that flows between endpoints can contain multiple proposals. Each proposal field is followed by a Transform Payload field, with the latter containing such information as the authentication method and encryption algorithm to be used. The opposite endpoint responds to the first message indicating which proposal, if any, it will support. If the initial message contains only one proposal, the opposite endpoint can simply acknowledge that the proposal is acceptable. Both the initial message and response to that message flow in the clear and are unauthenticated.

A second pair of messages is then exchanged after the first pair. This second set of messages results in the exchange of Diffie-Hellman public key random numbers that are used to establish a master key, referred to as a SKEYID. This pair of messages also flow in the clear and are not authenticated.

A third pair of messages is exchanged that contain information that enables each endpoint to authenticate the Diffie-Hellman exchange. This third pair of messages is protected both through authentication and by the use of an encryption algorithm based upon keying information exchanged by prior messages. Thus, a series of three pairs or six messages are exchanged between endpoints to initially establish a security association and exchange keys.

The actual method used for authentication under Phase One can be a shared secret, certificate or the use of public and private encryption keys. The use of a shared secret is not acceptable since a separate secret should be used for each communication. The use of certificates until recently has been minimal due to the slow pace associated with the integration of Certificate Authority (CA) systems into corporate environments. Because the use of public and private keys can be used in combination with a Diffie-Hellman exchange to both provide authentication and keying data, its use has increased.

Phase Two

Once the Phase One negotiation process is completed, the next phase requires another exchange of messages. These messages are used to define the security associations and keys that will protect IP datagrams.

There are two types of keys that can be used for ESP: symmetrical and asymmetrical. Symmetrical key encryption results in the same key used for both encryption and decryption and is used by a private key system. In comparison, asymmetrical key encryption results in the use of a pair of keys, one of which is published while the other remains secret or private, with the key pair related mathematically.

Because encryption is one-way with a private key used to decrypt data encrypted with the public key, two pairs of keys are required. Thus, the use of a key pair in each direction results in the encryption process using asymmetrical keys.

Under Phase Two an IPSec compliant endpoint will select a random value and transmit it to its opposite endpoint along with one or more proposed security associations, a key exchange payload that includes the prime number and generator used to generate public encryption keys and other data. Because Phase One negotiations resulted in the selection of a method to protect ISAKMP messages, the Phase Two messages are protected by the encryption algorithm agreed to during the prior phase. Thus, the Phase Two initial message is protected as it flows to the opposite endpoint. That endpoint responds under Phase Two with a random valve selected by the opposite endpoint. At this point in time each endpoint has enough information to derive a pair of keys, one for transmission and the second for reception of data. Although only two messages are required to exchange key information under Phase Two, a third message is transmitted by the initial endpoint that commenced the exchange. This message is transmitted by the initial endpoint to its opposite peer to indicate that the key exchange is correct. This third message contains a hash value that covers the random values previously exchanged, enabling the opposite endpoint to verify that all is well.

Key establishment

Under IPSec, key management must provide support for two key establishment methods: manual and automatic. Under manual keying an administrator becomes responsible for providing the keys and other initialization data required for security associations. As you might expect, manual keying is practical for small networking situations where there are a limited number of hosts and gateways. For example, consider an organization that has four locations that they wish to interconnect via the Internet. If each VPN relationship has a unique key, then the number of keys is $n*(n-1)/2$ or 6. Now suppose the organization wants to interconnect ten sites. The number of unique keys would be 45. Because each gateway must be manually configured and keys need to be shared with all corresponding gateways, manual keying can be both tedious and time-consuming. In comparison, automatic keying provides for the automatic deployment of keys. Although there is a degree of overhead associated with automatic key management, it removes the previous burdens associated with the manual keying process and is well suited for use by organizations with more than a handful of locations requiring interconnection.

The use of automatic key management can result in the creation of several keys for a security association. This potential action results from the fact that some encryption algorithms, such as public keys, require more than one key. In addition, some authentication algorithms can require more than one key while the process of re-keying to protect future transmission in the event a key is compromised will also result in multiple keys being required.

In concluding our discussion of encryption keys, it is important to note that automatic key management can provision keys via single or multiple strings. When there is only one key for a security association, a single string will suffice. However, when there are multiple keys, multiple strings are required to distribute the appropriate keys.

5.2 Working with IPSec

In this section we will turn our attention to the use of IPSec in a Microsoft Windows environment. At the time this book was prepared IPSec was supported on Windows XP Professional and Windows 2000 Server as well as the recently introduced Windows 2003 server. In this section we will primarily examine the configuration of what is normally the terminal point of an IPSec tunnel, which is the server. Thus, the primary focus of this section will be how we can create and manage IPSec policies in a Windows 2000 server environment.

5.2.1 Configuring IPSec policies

IPSec services are configured through the creation and application of IPSec policies. In a Windows environment IPSec policies can be stored in two locations: the Active Directory or the registry on a local computer. When IPSec policies are developed to satisfy the security requirements of a large number of employees affiliated with a domain, site, or organizational unit, storage normally occurs in the Active Directory. In comparison, the configuration of IPSec policies in a non-Active Directory environment can occur in the computer's registry.

Because the first step in configuring IPSec policies in a Windows 2000 server environment involves the use of the IP Security Policy Management snap-in, we will turn our attention to its use. However, prior to doing so it is worth noting that that snap-in is included in both Windows 2000 server and Windows XP Professional as an administrative interface that facilitates the creation and administration of security policies. That interface is obtained by adding the IPSec snap-in to the Microsoft Management Console (MMC). Similar to many other operations, to define IPSec policies on a computer you need to have appropriate administrator rights to Group Policy or be a member of the local system administrators group.

5.2.2 Adding the IPSec snap-in

The additional of the IPSec snap-in to the Microsoft Management Console (MMC) is a relatively straightforward process. You would first activate MMC via the Start menu, selecting the Run entry and entering MMC. The dialog box hiding in the background in the left portion of Figure 5.13 illustrates a portion of the resulting display of the MMC. Next, in the MMC you would

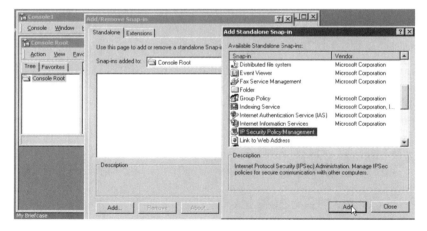

Figure 5.13   To initiate IP Security Policy Management, you must add that snap-in to the Microsoft Management Console.

click on the Console menu and select the Add/Remove Snap-in entry in the resulting menu.

The middle dialog box shown in Figure 5.13 illustrates the result of this action. Clicking on the button labeled 'Add' results in the display of available standalone snap-in modules. The dialog box shown in the right foreground of Figure 5.13 illustrates the available standalone snap-in modules on the Windows 2000 server used by this author. Note that the IP Security Policy Management snap-in is shown highlighted. Once that entry is highlighted, you would then click on the button labeled Add in that dialog box. This action would result in the display of a dialog box labeled 'Select Computer', which is illustrated in Figure 5.14.

In examining the entries in Figure 5.14, you would select the local computer entry if you want to manage IPSec policies only on the computer on which the MMC is running. If you want to manage IPSec policies for any domain

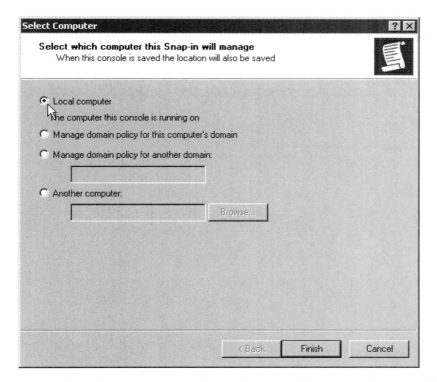

Figure 5.14   The Select Computer dialog box provides you with the ability to select which computer the previously selected snap-in will manage.

members, you would select the radio button associated with 'Manage domain policy for this computer's domain.' The other two entries, 'Manage domain policy for another domain' and 'another computer' provide you with the ability to manage IPSec policies for a domain of which the computer running the console is not a member or to manage a remote computer, respectively. Once you select the applicable computer, you would click on the button labeled 'Finish' and then click on the button labeled 'Close' on the Add Standalone Snap-in dialog box previously shown in the foreground in Figure 5.13.

We can view the results of the previously described series of actions to add an IPSec Policies capability to the MMC. Figure 5.15 illustrates the console showing IP Security Policies being placed on the local computer. By default there are three named entries for IPSec Policies as shown in the right window in the referenced figure. The first named entry, 'Client (Respond Only}' results in communications normally being in the clear or unsecured. This predefined policy is for computers, which should not secure communications most of the time, such as clients that only need IPSec when requested by another computer. This policy includes a default response rule, which permits the client to negotiate with computers requesting IPSec. Since the client responds

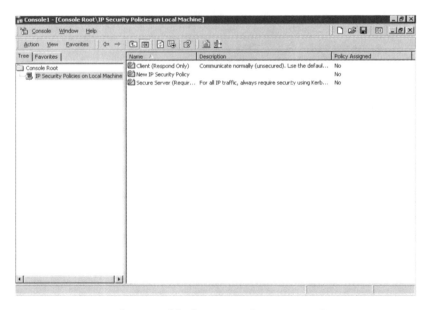

Figure 5.15 Viewing the establishment of IPSec policies on a Windows 2000 server.

to IPSec requests only, the requested protocol and port traffic defined by the requester are secured.

The second entry, 'New IP Security Policy' is used to directly create a new IPSec policy. The third entry, 'Secure Server (Require Security)' is used for computers that always require secure communications. By default, Kerberos is defined for authentication and the selection of this entry precludes the use of unsecured communications.

5.2.3 Creating an IPSec Policy

You can view the three entries in the right window of Figure 5.15 as templates that enable you to develop specific policies to satisfy your organization's operational requirements. In reality, the templates represent predefined policies you can use as-is, modify or ignore. To add or create a new policy you first select one of the three entries under Description and then select the Create IP Security Policy entry from the Action menu. Figure 5.16 illustrates the entries in the drop down Action menu, with the 'Create IP Security Policy' entry shown highlighted. Under Windows 2000 this action results in an IP Security Policy wizard being invoked. This wizard will enable you to specify the level

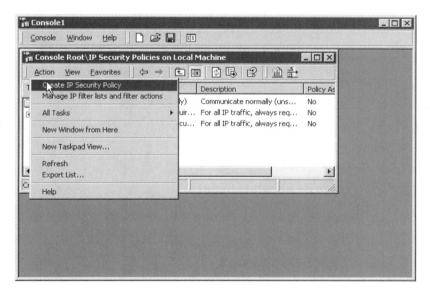

Figure 5.16 Using the action menu in the MMC to create an IP Security Policy.

of security to use when communicating with specific computers or groups of computers located on a network as well as for specific types of IP traffic.

5.2.3.1 Using the IPSec Policy Wizard

The IPSec Policy wizard represents a series of screen displays that facilitates the creation of a security policy. After the display of a welcome screen the wizard will prompt you to enter a name for the security policy and optionally provide a brief description. The description can be meaningful if you indicate its intention, such as which groups or domains it affects.

Responding to requests for secure communications
The next screen, which is shown in Figure 5.17, permits you to denote how the policy you are creating will respond to requests for secure communications. If you activate the default response rule by clicking on the checkbox, the computer will respond to remote computers that request security using the predefined default response rule. Although the term 'default' is used, in reality

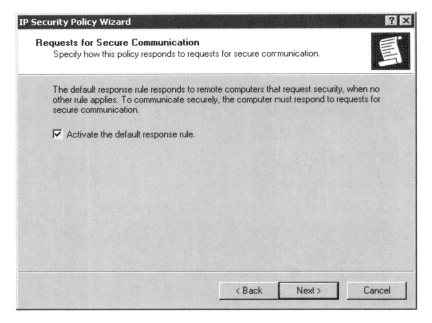

Figure 5.17 The IPSec policy wizard allows you to specify how the policy you are creating responds to requests for secure communications.

the term means predefined as Microsoft provides you with the ability to tailor the default, as we will shortly note.

Authentication methods

The IPSec Policy wizard under Windows 2000 by default uses the Kerberos V5 protocol for authentication. As an alternative, you can use a certificate from a certificate authority (CA) for initial authentication by clicking on a different radio button and selecting a particular certificate.

Figure 5.18 illustrates the display of the authentication screen by the IPSec policy wizard. Note that by selecting the lower radio button you can use a predefined string shared among computer to protect the key exchange, which corresponds to Phase One authentication that was described in the first section in this chapter. Thus, if your organization does not have a Kerberos server and you wish to avoid the complexity of periodically checking the status of certificates, you can enter a string that will be hashed and transmitted for authentication.

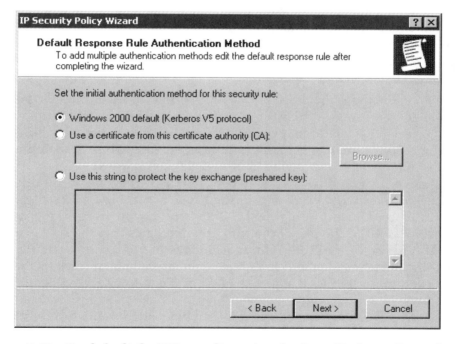

Figure 5.18   By default the IPSec policy wizard selects Kerberos for authentication.

Because this author's organization did not operate a Kerberos server and wished to obtain maximum flexibility, the string 'now is the time for all good men to come to the aid of their party and bring a six pack of diet coke' was used for authentication. To enter this string, the third radio button in Figure 5.18 was activated, enabling this author to enter the previously described string. Note that this author purposely lengthened the well-known string to minimize the possibility that an unauthorized third party could guess the string. Because the authentication key value (string) is stored in the IPSec policy in a readable format, this means that any user that is authenticated can read the string, although its value can only be changed by the directory administrator. Thus, the use of an authentication key in a directory based policy can represent a potential security problem and you are better off using authentication strings when an IPSec policy is stored and managed on the computer to which it applies.

Once you select an applicable authentication method and click on the button labeled 'Next,' the wizard will display a completion box. That dialog box will include an Edit Properties checkbox whose default setting is selected. Thus, if you leave the default setting as is and close the wizard you will immediately have the ability to edit the properties of the IPSec policy you just created. Thus, moving forward, the next logical step is to examine how a policy can be modified or edited.

5.2.3.2 Editing IPSec Policy Properties

Figure 5.19 illustrates the IPSec policies Properties dialog box displayed if we double click on the previously created policy or when we close the wizard after accepting the default that we wish to edit the policy. Because this author named the policy 'remote access,' the title of the dialog box in the upper left corner of Figure 5.19 is shown as 'remote access properties'.

To remove any potential confusion between the Terms policy and rules, the IP Security Rules window shown in the Rules tab in Figure 5.19 displays a list of rules contained in a particular IPSec policy you are working with. If a checkbox is set to the left of the rule, this indicates that the rule is active. Through the use of the Add, Edit and Remove buttons you can tailor rules for a policy. Every policy will have one or more predefined rules that cannot be removed, indicated by the gray shading of the button labeled 'Remove' in Figure 5.19.

Since we previously created the IPSec policy called 'remote access' using default settings, the window labeled 'IP Security Rules' only has one entry. That entry or any other rules in the IPSec policy can be edited or removed by selecting the rule and clicking on an applicable button.

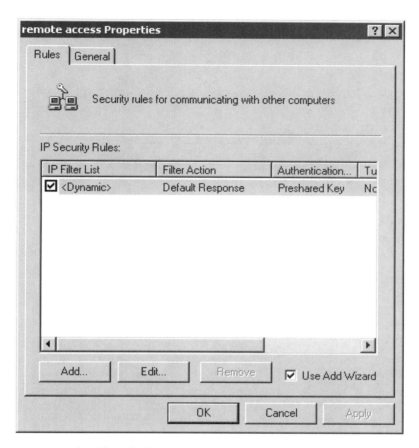

Figure 5.19 By double clicking on a previously created policy or closing the wizard with its default edit checkbox marked permits you to edit the policy rules.

You can actually either double click on a rule or select a rule and click on the button labeled 'Edit' to edit the previously selected rule. Continuing our work on the previously created rule, let us edit that rule to obtain a better understanding of the built-in capabilities of Windows 2000 with respect to configuring IPSec.

Security Methods tab
Figure 5.20 illustrates the Edit Rule Properties dialog box with its Security Methods tab placed in the foreground. Note the window provides a list of security methods to be used when negotiating security levels for communicating

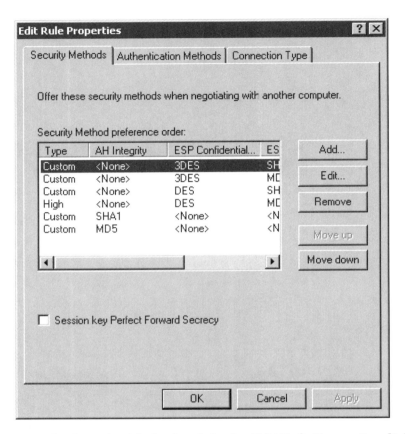

Figure 5.20 The Security Methods tab in the Edit Rule Properties dialog box permits you to define the security methods for negotiation when communicating with another computer.

with another computer. The list in the window is displayed in descending order of preference. Clicking on a Security Method is equivalent to selecting Edit, providing you with the ability to modify the selected security method. To obtain an appreciation for options supported by Microsoft, this author selected the first entry in the window, which is shown highlighted in Figure 5.20.

The result of double clicking on the first security method or selecting the button labeled Edit in Figure 5.20 when the first method is highlighted allows us to Edit that method. Editing a security method permits us to select either ESP, AH or a custom setting, after which the selection of a custom setting permits tailoring by being able to specify the integrity algorithm and

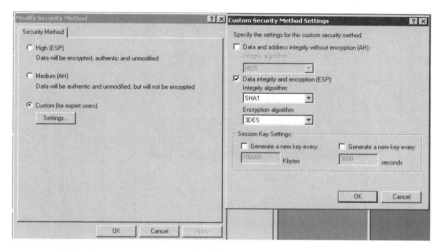

Figure 5.21   Editing a security method and examining custom security method settings.

encryption algorithm as well as when a new key should be generated for the encryption algorithm. To illustrate the preceding let us double click on the first entry in Figure 5.20 and examine the custom settings that are available for selection.

The left portion of Figure 5.21 illustrates the Modify Security Method dialog box with its custom entry selected. Note you could simply select ESP or AH and avoid the necessity of custom settings.

The right side of Figure 5.21 illustrates the default custom settings. Note that by default data integrity and encryption (ESP) is selected and the SHA1 algorithm is used as the integrity algorithm. SHA1 uses a 160-bit key and is the strongest algorithm for integrity. Other options include MD5, which uses a 128-bit key and 'none,' with the latter to be used if you selected the first option 'Data and Address integrity without encryption (AH).'

Currently Microsoft supports two encryption algorithms, 3DES, which is the default setting and provides the highest level of confidentiality compared to the other setting, which is DES. You can also configure a custom encryption setting by selecting 'none' if you want the ESP format but do not require privacy. At the bottom of the Custom Security Method Settings dialog box is a session key setting area. You can use either checkbox to select an interval, either in kilobytes or seconds, after which a new key will be generated. Clicking on either session key setting checkbox provides you with the ability to either accept the default value or enter a new session key setting.

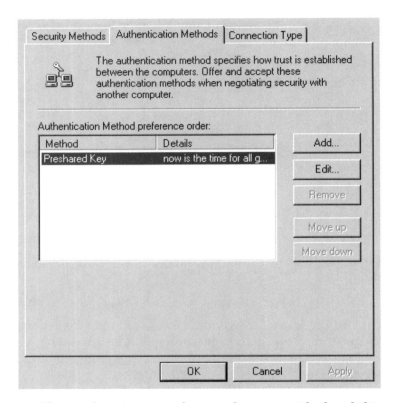

Figure 5.22 The Authentication tab provides you with the ability to define or revise how trust is established between computers.

Authentication tab

Returning to the Edit Rule Properties dialog box shown in Figure 5.20, the second tab that is labeled 'Authentication' provides you with the ability to examine and, if necessary, change the manner by which trust is established between computers. Figure 5.22 illustrates the setting of the Authentication Methods Tab, which shows a portion of the pre-shared key we previously entered. If you click on the button labeled Edit, you can change the string or select the use of Kerberos or a certificate, with the resulting dialog box appearing similar to a portion of the IPSec Policy wizard previously shown in Figure 5.18.

Connection Type tab

The third tab in the Edit Rule Properties dialog box is labeled Connection Type. This tab provides you with the ability to associate a network connection with the rule you established. There are three options on the Connection Type tab: 'All Network Connections' which is the default as well as 'Local Area Network' and 'Remote Access.' Selecting the 'All Network Connections' option results in the rule being applied to all network connections you created on the computer. In contrast, the other two options restrict the rule to either all LAN connections or remote connections.

General tab

The second tab in the selected IPSec Properties dialog box is labeled 'General.' The selection of this tab provides you with the ability to view and, if necessary, modify general IPSec properties. Those general IPSec properties include having the ability to examine default key exchange properties and, if necessary, modify those settings. Once again, we will use a sequence of three dialog boxes in one illustration to examine the options available for selection.

The left dialog box positioned in the background of Figure 5.23 illustrates the display of the General tab from the remote access properties dialog box, while the middle portion of the display shows the initial Key Exchange Settings dialog box resulting from clicking on the button labeled Advanced in

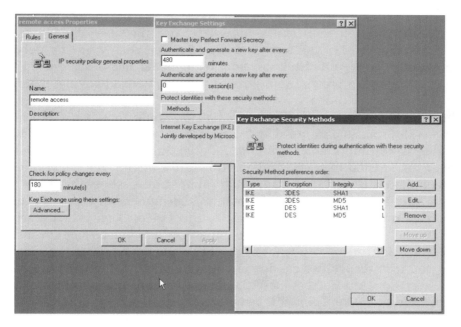

Figure 5.23 The General tab provides the ability to examine and, if necessary, alter key exchange settings.

the lower portion of the General tab. The dialog box located in the foreground in the right portion of Figure 5.23 results from clicking on the button labeled 'Methods' located in the middle dialog box. Thus, Figure 5.23 illustrates the sequence of dialog boxes that can be displayed when you are working with the General tab in an IPSec Properties dialog box.

If we first focus our attention upon the General tab in the dialog box in the left portion of Figure 5.23, we note the display of the name we previously assigned as well as a window that permits us to enter a description of the policy. That description can be up to 255 characters and should be used to identify why this policy was created, the computers the policy is applicable to, and similar information. If your organization is simply creating one policy for all remote users, then the name of the policy used by this author could be sufficient and you might then elect not to enter a description.

At the lower left portion of the General tab you will note a box that shows a default value of 180. That value represents how often, in minutes, the Active Directory will be polled for changes to the policy and the value is only applicable if you are configuring a policy for a computer that is a domain member. Under the policy change box is a button labeled 'Advanced'. Clicking on that button results in the display of the Key Exchange Settings dialog box, which is shown in the middle portion of Figure 5.23. That dialog box provides you with the ability to view and, if necessary, alter the key exchange settings.

The first checkbox in the Key Exchange Settings dialog box by default is not checked. You would check this box, which is labeled 'Master key Perfect Forward Secrecy' to prevent the reuse of previously used keying material or keys to generate additional master keys.

The second box in the Key Exchange Settings dialog box is used to specify the interval, in minutes, after which a new key will be generated. By default the value is 480 minutes or 8 hours. The next box, whose default value is shown as 0, is used to specify when authentication and the generation of a new key should occur, based upon the number of sessions that have taken place. The purpose of this box is to limit the number of times the master key can be reused as keying material for the session key. When a security association is established with other computers, the same master key material can be reused. Thus, if you want to limit the number of times the reuse can occur, you would specify a session key limit. Note that if you enable 'Master key Perfect Forward Secrecy' any parameter entered into the 'Authenticate and generate a new key' box will be ignored.

Turning our attention to the third dialog box, which is shown in the foreground in the lower right portion of Figure 5.23, you would click on the button labeled 'Methods' in the second dialog box to display the 'Key Exchange Setting Methods' dialog box. This dialog box provides you with the ability to view and, if required, modify the default configured security methods that will be used for the key exchange. If you look at the dialog box labeled 'Key Exchange Security Methods' shown in the foreground of Figure 5.23, you will note that by default four security methods are listed in preference order for protecting identities during authentication. The first three security method columns that are visible indicate the type, encryption and integrity. The pre-defined type for all entries is the IPSec Internet Key Exchange (IKE), while encryption options are 3DES and DES. Integrity options include the 160-bit SHA1 hash and the less secure 128-bit MD5 hash. Mostly obscured in the Security Method preference order is a fourth column, which is labeled 'Diffie-Hellman'. That column indicates the length of base keying material used to generate actual keys. There are two types of entries that can occur in the fourth column: Low (1) and Medium (2). Low results in the use of 768 bits as the length of base keying material to generate the actual keys, while Medium results in the use of 1024 bits as a basis and results in a stronger key generation process. The first two rows in the Security Method preference order window are pre-defined for Medium (2), while the last two rows are defined for Low (1). Because SHA uses a 160-bit value in comparison to the 128-bit hash used by MD5, and 3DES is obviously stronger than DES, it is relatively easy to verify that the Security Method preference order varies from the strongest to the weakest.

The security methods listed in the preference order function as mechanisms to protect identities during the initial authentication and key exchange process. The list is ordered in precedence from strongest to weakest and defines which security method, key settings and algorithms will be used to provide protection. For most organizations the default settings that provide the strongest protection should be sufficient.

By double clicking on a security method or selecting a method and clicking on the Edit button you can change the integrity algorithm, encryption method or the Diffie-Hellman group for a particular entry, although as previously discussed, the default settings should be acceptable for most organizations.

After you complete working with the General tab and its dialog boxes for the policy and any rules you just created, you need to consider placing filters in either client or server policies as a further mechanism to secure your IPSec policy. Last, but not least, once you complete the policy, you will need to assign it so that it goes into effect. In the remainder of this section we will discuss both tasks.

5.2.4 Working with IPSec filters

The purpose of IPSec filters is to enable you to associate IP addresses and protocols with an IPSec policy. Doing so enables you to better tailor a previously created policy. In order to examine how we can work with IPSec filters, we must first select a policy.

5.2.4.1 Adding or Editing IPSec Filters

To add or edit IPSec filters you would right-click on a policy you wish to modify in the IPSec Policy Management box and click on the Properties entry. Figure 5.24 illustrates the selection of the previously created remote access IPSec policy by right clicking on the entry. This action results in the display of a pop-up menu in which the Properties entry is shown selected in the previously referenced illustration.

Once you select the Properties entry in the pop-up menu, the Properties dialog box for the policy you wish to modify will be displayed. This dialog box, which is shown in the left portion of Figure 5.25 has two tabs: 'Rules' and 'General' similar to Figure 5.19 with the Rules tab positioned in the foreground. The Rules tab window lists IP filters you can modify by clicking

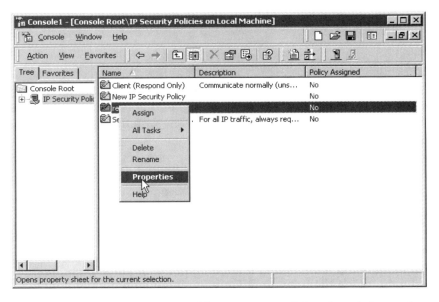

Figure 5.24 To add or edit IPSec filters you would right click on the policy you wish to modify and select the Properties entry from the pop-up menu.

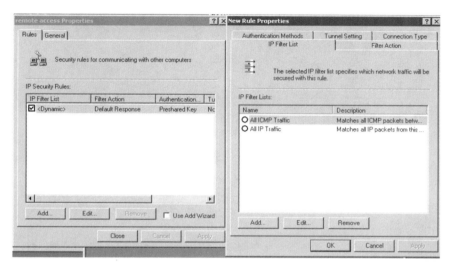

Figure 5.25   By default IP filters are available for all ICMP traffic and all IP traffic.

on the Edit button. In examining the IP Filter list on the Rules tab, note the box in the lower right of the display marked Use Add Wizard. By default that box is checked and clicking on the button labeled 'Add' results in the use of a Security Rule Wizard that guides you in defining the source destination, and type of IP traffic that will be associated with a security rule. In addition, the wizard will enable you to define IP tunneling attributes, authentication methods and filter actions. By clearing the check box you can create a filter manually by clicking on the button labeled Add. As a third option, you can use the Rules tab to reconfigure an existing filter by clearing the check box and clicking on the button labeled Edit. Because the manual process provides a more concise visual indication of options, we will primarily focus our attention upon this method. However, we will also look at an applicable wizard display to note the similarities and differences between a manual configuration of filters and the use of the wizard.

The IP Filter list tab

Turning our attention to the right portion of Figure 5.25, note that the New Rules Properties dialog box contains five tabs, with the IP Filter list tab by default displayed in the foreground. From the IP Filter list you can select one of two predefined lists – all ICMP traffic or all IP traffic – or you can create a new IP filter list by clicking on the button labeled Add. Because simply

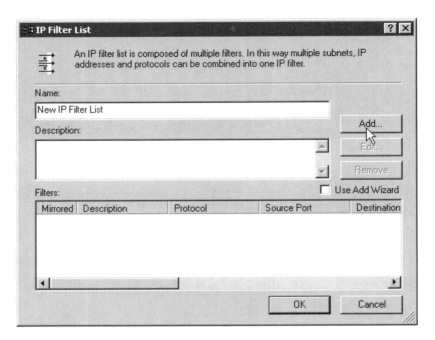

Figure 5.26   The IP Filter list provides a mechanism to associate source and destination IP addresses and a protocol to a filter.

selecting all ICMP traffic or all IP traffic will rarely satisfy one's operational requirements, let us create a new IP filter list to specify which network traffic will be secured. To do so we would click on the button labeled 'Add' in the New Rule Properties dialog box shown in the right portion of Figure 5.25, with the IP Filter list tab in the foreground. This action results in the display of the IP Filter list box shown in Figure 5.26.

The IP Filter list dialog box provides you with the ability to name the filter, provide a text description for the filter, and specify one or more filters in the form of source and destination IP addresses or subnets, and protocols, as well as enter an explanation of the function of the filter. As we will shortly note, both the manual process as well as the use of the wizard have similarities, although the manual process provides the user with a more concise configuration capability.

Filter Properties
From the IP Filter list dialog box shown in Figure 5.26 the selection of the Add button provides you with the ability to specify filter properties to include

source and destination addresses, protocol, and a description of the filter. Since the motto of the state of Missouri is 'show me,' let us pretend we are natives of that great state and cycle through the screens necessary to create a filter.

Addressing operations

Clicking on the Add button results in the display of the Filter Properties dialog box, which is shown in Figure 5.27. Note that this dialog box has three tabs, with the Addressing tab by default positioned in the foreground. The Source Address menu is shown pulled down to illustrate the options available for selection. The same options are also available for the destination

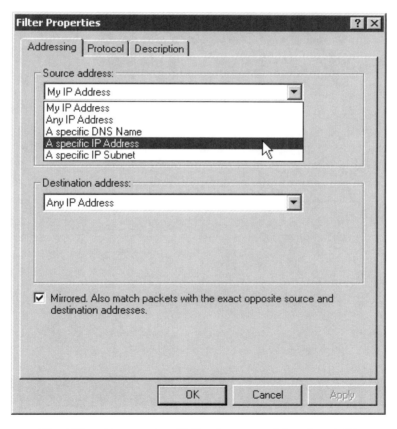

Figure 5.27   The Filter Properties dialog box provides the ability to associate source and destination IP addresses and a protocol of the filter.

IP address. In the lower left corner of the Filter Properties dialog box you will note a check box, which is set by default. This check box, which is labeled 'Mirrored,' provides an automatic packet match using the opposite source and destination addresses specified to provide a bi-directional filtering capability when checked.

If you specify specific IP addresses other than my IP address, the display of the addressing tab will change. An example of this change is shown in Figure 5.28. In this example, this author is creating a filter from the specific source IP address of 198.78.46.8 to the destination IP address of 205.131.175.6. Because the default Mirrored check box is enabled, this filter will also be applicable to packets flowing from the IP address of 205.131.176.6 to the IP address of 198.78.46.8.

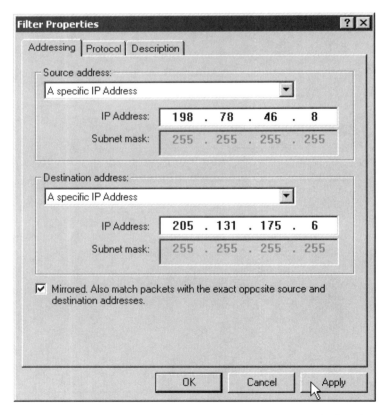

Figure 5.28 The Addressing tab in the Filter Properties dialog box provides the ability to assign source and destination addresses to the filter.

To enable the settings shown in Figure 5.28 you would click on the Apply button. Assuming you did so, let us bring the Protocol tab to the foreground and examine its settings.

Protocol selection

Figure 5.29 illustrates the Filter Properties dialog box with its Protocol tab positioned in the foreground. Protocols you can select range from 'Any' to EGP, ICMP, TCP, UDP, XNS-IDP and several minor protocols hardly used. When you select a protocol, its numeric identifier is displayed below the selected protocol. For certain protocols, such as TCP and UDP, their selection enables you to set the source (from) and destination (to) IP protocol ports.

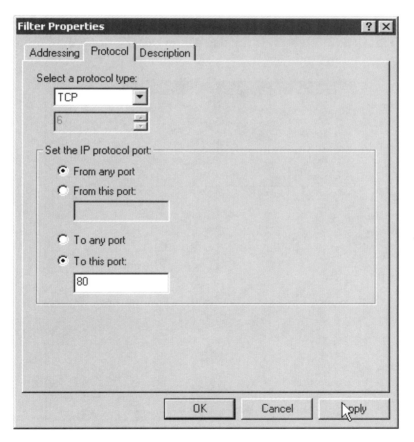

Figure 5.29   The Protocol tab in the Filter Properties dialog box provides the ability to select an IP protocol and source and destination ports for the filter.

Note that in Figure 5.29 the window labeled 'Select a protocol type' is shown set to TCP. This action resulted from the selection of TCP from a drop down menu of protocols. Also note that because the destination port defines the application, this author set the IP protocol port such that an IP datagram can have any source port value but must be set to 80 for the destination port. This ensures that Web services from the previously defined source IP address configured in Figure 5.27 to the destination IP address configured in that illustration will be set in the filter we are creating. Similar to other dialog boxes, you need to click on the button labeled 'Apply' to Save the protocol settings.

Description tab

The third tab in the Filter Properties dialog box, which is labeled 'Description,' provides you with the ability to enter a description about the filter you are creating. Figure 5.30 illustrates the use of the Description tab to enter the text

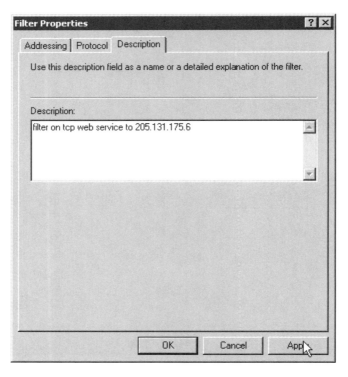

Figure 5.30   The Description tab in the Filter Properties dialog box provides a mechanism to explain the function of the filter.

string 'filter on TCP web service to 205.131.175.6' which basically defines the filter we just created. Once again, you need to click on the button labeled 'Apply' to make any changes previously entered take effect.

Now that we have used all three tabs in the Filter Properties dialog box let us close the box by clicking on the button labeled OK and return to the IP Filter list dialog box.

If we compare the IP Filter list box previously shown in Figure 5.26 to the box shown in Figure 5.31, we can note the effect of our recent series of filter operations. In the window labeled 'Filters' in the lower portion of Figure 5.31, you will note the horizontal entry of the filter we just created. By moving the horizontal scroll bar to the right you can view additional summary information about the filter that was just created. Although we used the Description tab in the Filter Properties dialog box to describe the filter, it is possible to use the IP Filter list dialog box to create several filters. Thus, the window labeled 'Description' provides you with an additional opportunity to describe the filter or a series of filters. For illustrative purposes we are only creating a single filter and will use the Description window to enter the text 'PROTECT WEB ACCESS' to further define the filter.

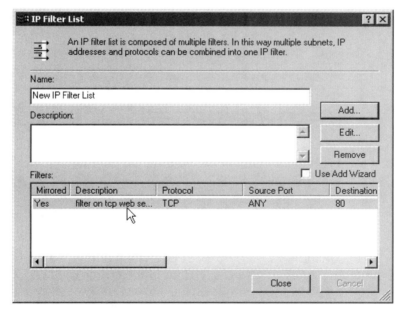

Figure 5.31 The newly created filter has its parameters indicated in the Filters window in the IP Filter list dialog box.

5.2.4.2 Comparison to Using the Wizard

If you look at either Figures 5.26 or 5.31 you will note that the check box associated with Use Add Wizard was not checked, resulting in the manual creation of a filter. If you accept the default check box setting and use the Filter Wizard, instead of being able to enter multiple information on a dialog box, you will be gradually walked through the entry of information. For example, Figure 5.32 illustrates the use of the IP Traffic Source display when you use the wizard. If you use the wizard you will need to work with multiple screens to basically obtain the same results as when you use the manual configuration method. This can be observed by noting that the Addressing tab in the Filters Properties dialog box shown in Figure 5.27 provides you with the ability to enter both source and destination addresses. In comparison, the use of the wizard requires you to be walked through four separate screens to obtain the same end result. Thus, most people should be able to speed up the configuration process by manually configuring IP filters.

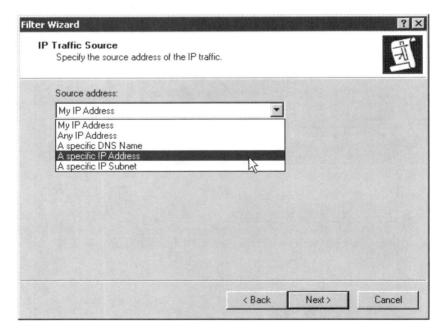

Figure 5.32   The Filter Wizard walks you through the use of single data entry screens that show the configuration process.

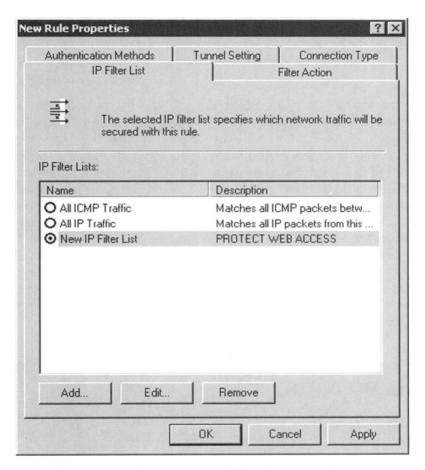

Figure 5.33 The New Rule Properties dialog box permits you to associate various security-related measures with the selected filter.

5.2.4.3 Working with Filter Properties

After you complete your filter list, Windows will return you to the display of the New Rule Properties dialog box, similar to the one shown in Figure 5.33.

Note that the filter list titled 'New IP Filter list' represents the list we have just created to protect web access and is highlighted in the IP Filter list windows. This action ensures that through the use of the other tabs in the dialog box we can customize rules for the newly created filter.

As we work our way through the four additional tabs in the New Rule Properties dialog box we will note the similarities of some options to previously discussed IPSec settings.

Filter Action tab

The first tab in the New Rule Properties dialog box we will examine is the Filter Action tab. This tab, which is shown moved to the foreground in the left portion of Figure 5.34, allows you to specify how the filter rule will be employed to secure traffic. As indicated in the left portion of Figure 5.34, there are three predefined filter actions from which you can select. The first action, 'Permit,' allows unsecured IP packets to pass through. The second action, 'Request Security (optional),' accepts unsecured communications, but requests clients to respond using IPSec. This option enables communication with non-IPSec-aware computers. In comparison, the third option, that is labeled 'Require Security' and shown selected, accepts unsecured communication, but always requires clients to respond using IPSec. By double clicking on each option you will display a Require Security Properties dialog box that is shown in the right portion of Figure 5.34 for the filter selected in the left window.

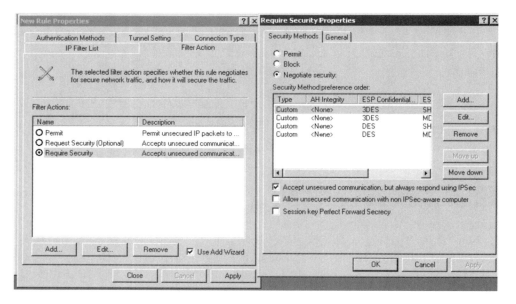

Figure 5.34 The Filter Action tab can be used to associate a specific security method with a previously defined filter action.

The right portion of Figure 5.34 illustrates the Security Methods tab positioned in the foreground of that dialog box. Note that the Security Method preference order is similar to the same window in the Edit Rules Properties dialog box previously shown in Figure 5.20. However, the Security Methods tab in the Require Security Properties dialog box provides you with the ability to specify how the security method will be applied to the filter as well as the security method preference order. Thus, the Security Method tab allows you to permit, block, or negotiate security using the security method preference order shown in the window in the right portion of Figure 5.34.

From the Require Security Properties dialog box shown in the right portion of Figure 5.34 you can click on the Add or Edit buttons to select an applicable security method. The selection of either button will result in the display of a Security Method dialog box similar to the one shown in the left portion of Figure 5.35. If you clicked on the Add button, the dialog box will be titled 'New Security Method.' In comparison, if you clicked on the Edit button, the dialog box will be titled 'Modify Security Method' as shown in the left portion of Figure 5.35.

Prior to discussing the entries in the left side of Figure 5.35, note that by clicking on the Settings button the Custom Settings dialog box is displayed,

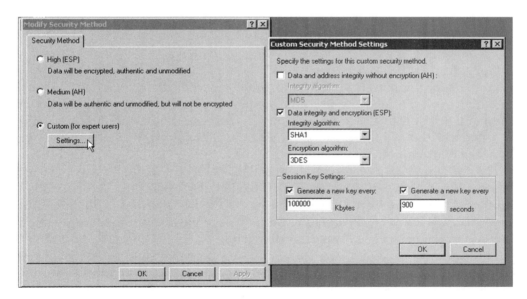

Figure 5.35 When working with a filter certain default values associated with key settings change from their policy default values.

which is shown in the right portion of the illustration. If you compare Figure 5.35 to Figure 5.21 you will note that although both illustrations are similar, when you compare the default custom settings, you will notice some slight differences between the two. That is, when we first configured IPSec via the use of the wizard we created the dynamic IP Filter list entry shown in Figure 5.19. When we edited the rule, the default custom settings defined the use of SHA1 for the integrity algorithm and the use of 3DES as the encryption algorithm similar to the custom settings shown in the right portion of Figure 5.35. What changed is the fact that when we created our filter the Session Key settings boxes are now set to generate a new key either every 100,000 kilobytes or 900 seconds. Note that in addition to the session key settings now being enabled, the default time for generating a new key is reduced from 3600 to 900 seconds.

If we return to the right portion of Figure 5.34 we note that there are two tabs in the dialog box, with the tab labeled 'General' in the background. The General tab simply defines the action you selected in the check boxes in the lower portion of the Security Methods tab. To illustrate this, the General tab is positioned in the foreground and is shown in Figure 5.36.

Because we selected the Require Security filter, the box associated with 'Accept unsecured communications but always respond using IPSec' was checked in the lower right portion of Figure 5.34. Thus, the General tab provides a more detailed description of the security properties configured for the filter with which we are working.

Other tabs

As indicated in Figure 5.33, the New Rule Properties dialog box has five tabs, of which we examined the IP Filter list tab. In concluding our discussion of the configuration of IPSec we will briefly discuss the use of the other tabs in the dialog box as their use is the same as tabs or wizard dialog boxes we previously examined.

The Authentication Methods tab is used to specify the manner by which trust is established between computers. This tab is the same as the 'Authentication Methods' tab in the Edit Rule Properties dialog box previously illustrated in Figure 5.22. However, by default, Kerberos is shown as the authentication method in the New Rule Properties dialog box.

The 'Connection Type' tab functions as previously described when we discussed the 'Edit Rule Properties' dialog box. That is, you would use this tab to select the network connection from the three options: All Network Connections, Local Area Network and Remote Access.

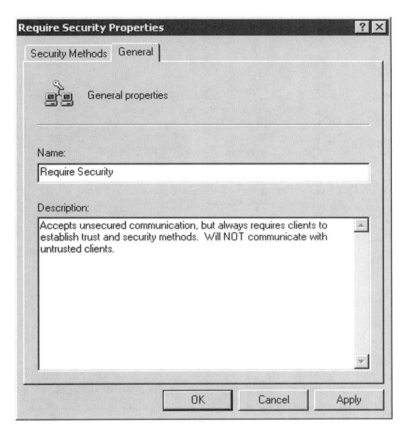

Figure 5.36 The General tab provides a more detailed description of the security properties configured for the filter with which you are working.

The Tunnel Setting tab, which is illustrated in Figure 5.37, by default is not used for the rule you are creating.

As a refresher, a tunnel endpoint represents the tunneling computer closest to the IP traffic destination. If you enable a tunnel endpoint, communications will only be exchanged with a specific computer, which would not be our intent if we were creating a rule for many remote computers to securely access a single server.

In concluding our review of the tabs in the New Rules Properties dialog box, we need to discuss the Filter Action tab. That tab is used to define how the filter rule will negotiate for secure networks traffic and how it will secure

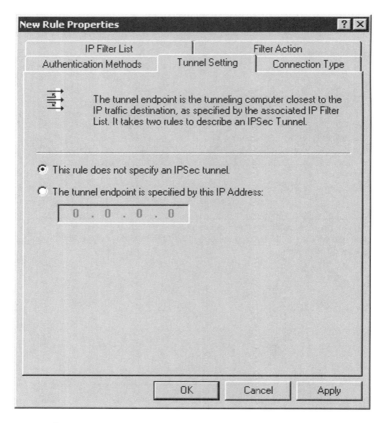

Figure 5.37   Under Microsoft terminology the use of a tunnel results in communications being exchanged with a specific computer.

such traffic. This tab is the same as the Filter Action tab shown in the left portion of Figure 5.34 for the New Rule Properties dialog box.

5.2.4.4 Activating a Policy

One of the items often overlooked in establishing an IPSec policy is the failure to activate the policy. To activate a previously created policy you would right click on the policy to bring up a pop-up menu. From that menu you would select the assign entry, which will activate the previously created policy. Unfortunately, there is no Microsoft reminder to activate a policy, so if you fail to do so you may have wasted a considerable amount of effort as well as good intentions.

5.3 SSL and TLS

The Secure Sockets Layer (SSL) protocol, along with its successor, referred to as the Transport Layer Security (TLS) protocol, should be very familiar to most readers. SSL dates back to the efforts of the Netscape Communications Corporation during the early 1990s as a mechanism to provide Web security. In May 1996, development of SSL became the responsibility of the Internet Engineering Task Force (IETF). Because Netscape had been intimately involved in the original development of SSL, although other organizations assisted in the development of later versions of the protocol, the IETF renamed the protocol Transport Layer Security to negate the appearance of any bias towards a single vendor. The TLS specification was officially released in January 1999. Although the name of the protocol was changed, the differences between the two protocols for the most part are very minor. Today, SSL and TLS provide tens of millions of Web browser users with the ability to perform electronic business activities in a secure manner. Although many readers primarily associate SSL and TLS with Web browsing, SSL and TLS provide support for multiple applications. In this section we will first gain an idea of the rationale for SSL and the manner by which this protocol operates. Once this is accomplished we will turn our attention to the use of one of several popular network appliances that can be used to create SSL/TLS-based VPNs.

In the remainder of this section we will collectively refer to SSL and TLS through the use of the initials for the original protocol since they are very similar as, in addition, most people are more familiar with a reference to the former. Only when there are significant differences between the two that warrant attention will further reference to TLS be made in this section. That said, let us turn our attention to obtaining a basic understanding of SSL by first focusing on the rationale for its existence.

5.3.1 Rationale for SSL

Until the development of the Web browser most client–server communications that required security occurred over leased lines or via the dial network to fixed corporate locations. Security was commonly provided through the use of datacryptor type encryption devices that used the DES algorithm. This action required devices at the client and server to be configured with the same key.

In the pre-Web corporate communications world the distribution of DES keys and their periodic change were difficult as each remote user had to be contacted and provided with a new key. While this action was not too difficult to support a small number of regional offices that required secure communications with a central site or for a handful of dial users accessing one or a few locations, the Web changed the potential logistics of key distribution. Instead of a handful of clients or client locations requiring secure communications to a common location, clients could now number tens of millions and require secure access to randomly selected servers. Thus, the conventional delivery of keys to support symmetric encryption represented an impossible task.

Recognizing this problem, Netscape communications began its attempt to develop a flexible communications security protocol. This protocol would consist of messages and rules for the exchange of messages that would negotiate various security-related activities between clients and servers. By providing a mechanism to select authentication, encryption and other security parameters, the protocol would provide flexibility to support a variety of security-related techniques, some of which could be developed in the future. Thus, SSL can be viewed as a flexible communications protocol developed to negotiate and place into effect various security settings between clients and servers.

5.3.2 Overview of SSL

SSL can be viewed as a separate protocol layer that provides security for applications operating above that layer. In this section we will turn our attention to the position of SSL in the TCP/IP protocol stack and obtain a brief introduction to SSL network appliances, with detailed information about the latter covered later.

Figure 5.38 compares the TCP/IP protocol stack's HyperText Transport Protocol (HTTP) used to transport Web traffic in both secure and non-secure modes. In the secure mode an HTTP application interfaces with SSL, which is then transported by TCP at layer 4 in the protocol stack. In comparison, when HTTP is in a non-secure mode it is directly transported by TCP.

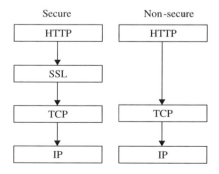

Figure 5.38   SSL represents a separate protocol layer, which provides security.

5.3.2.1 Need for Network Appliances

Although HTTP is shown at the application layer in Figure 5.38, it is also possible for SSL to be used to provide security for other applications. For example, SSL could be used to provide security for FTP and other Internet TCP/IP-based applications with appropriate interfaces from the application to SSL. Because the primary interface to SSL was developed for HTTP, several vendors developed network 'appliances' that support the use of SSL through a Web browser to access other applications at a central site where the appliance is located.

Although there is no appliance standard, each device essentially operates by functioning as a proxy between a client operating a browser using SSL and a series of TCP/IP applications. For example, assume an FTP server is located behind the network appliance as illustrated in Figure 5.39. A client would connect to the network appliance securely using SSL. After entering a user ID/password or token and PIN, or another authentication method that is acceptable to the appliance, that device will display a menu of applications it supports. Selecting an application results in the network appliance contacting the computer supporting the application on behalf of the Web browser client.

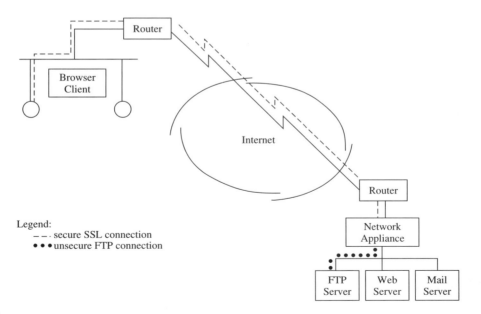

Figure 5.39 Using a network appliance.

In the example shown in Figure 5.39 the network appliance contacts the FTP server on behalf of the client, converting the HTTP protocol to the FTP protocol and vice versa in the opposite direction. Because most SSL network appliances support a large number of applications, clients can now access those applications through a browser, with security to the appliance provided through the use of SSL. Thus, there is no need to have SSL integrated into other applications.

As we will note later in this section, a key limitation on the use of such appliances to create a VPN is the fact that SSL requires the use of a reliable transport protocol, such as TCP. Thus, SSL cannot be used with UDP, which represents a best-effort or connection-less protocol.

5.3.3 SSL operation

As briefly mentioned earlier, SSL consists of a series of messages and rules governing the exchange of messages. Because SSL is designed to support client–server operations rules also govern who initiates communications and their actions during the security negotiation process. As you might surmise, clients are responsible for initiating a secure communications session. Since the client is responsible for initiating communications, it is also responsible for proposing a set of SSL security options to the server. The server, however, is responsible for selecting one of the security options proposed by the client that it wishes to support.

5.3.3.1 SSL Messages

The key to SSL client–server communications is a core set of messages that are exchanged between the two devices. Table 5.2 lists thirteen key SSL messages in alphabetical order as well as a brief description of the function of each message.

To obtain an understanding of the relationship between these messages let us examine their interaction when SSL is used to authenticate the identity of a server and to establish encrypted communications. Although SSL messages can be used to provide mutual authentication, in the wonderful world of e-commerce it is the client that primarily needs to authenticate the identity of the server. This is because typically the client will submit a credit card number that the server can verify and not the other way around.

5.3.4 Message exchange

SSL supports several types of message exchanges. In this section we will examine the sequence of messages required for two popular operations. First,

TABLE 5.2 SSL messages

Message	Function
Alert	Informs the other party of a potential security breach or communications failure
ApplicationData	Transports information exchanged between two parties
Certificate	Transports the sender's public key certificate
CertificateRequest	Sent by server to request the client provide its public key certificate
CertificateVerify	Sent by client to verify that it knows the private key that corresponds to its public key certificate
ChangeCipherSpec	Informs the other party to begin using an agreed upon security service, such as encryption
ClientHello	Transmitted by the client to indicate the security services it desires and those it supports
ClientKeyExchange	Transports client cryptographic keys to be used for secure communications
Finished	Indicates the completion of initial negotiations and that secure communications was established
HelloRequest	Transmitted by the server to request the client to start the SSL negotiation process
ServerHello	Transmitted by the server to indicate the security services to be used for communications
ServerHelloDone	Transmitted by the server to indicate it completed all of its requests to the client for establishing communications
ServerKeyExchange	Transmitted by the server, this message conveys cryptographic keys that will be used for communications

we will examine messages exchanged to enable a client to authenticate a server. Once this is accomplished, we will look at the sequence of messages exchanged to establish encrypted communications.

5.3.4.1 Authenticating a Server

As previously discussed, in an e-commerce environment clients primarily need to authenticate servers. Although it is possible to reverse authentication under SSL, most implementations of the technology are based upon the client authenticating the server. Thus, in this section we will first examine the SSL message exchange required to authenticate a server

Figure 5.40 illustrates the exchange of a series of SSL messages used for a client to authenticate the identity of a server. As you might expect, each message includes a series of fields that contain various parameters that provide

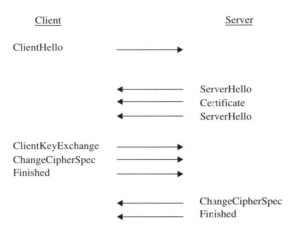

Figure 5.40 The sequence of SSL messages required to authenticate a server.

functionality to the message. For example, the Client Hello message, which is used to indicate the security services the client desires and those it is capable of supporting, includes both Session ID and CipherSuites fields. The session ID field identifies the SSL session while the CipherSuites field includes a list of cryptographic parameters that the client can support. In addition, the Client Hello includes a 32-byte Random Number field, which is used as a seed for cryptographic computations.

The server responds to the Client Hello message with a Server Hello. The Server Hello message selects SSL options to be used. Because the Client Hello included a request for authentication, the server follows its Server Hello message with a Certificate Message and a Server Hello Done message. The Certificate message includes a series of fields that include the server's public key certificate as well as the certificate authority's root certificate. The information provided in the Certificate message enables the client to verify the identity of the server. Once the Certificate message is transmitted, the server concludes its portion of the negotiation with a Server Hello Done message.

The client responds to the server with a sequence of three messages. First, the client transmits a Client Key Exchange message. This message includes session key information that is encrypted using the server's public key, which was previously received in the Certificate message transmitted by the server. Next, the client transmits a Change Cipher Spec message. This message informs the server to begin using the negotiated options, such as encryption, for all subsequent messages. The client then issues a Finished message, which informs the server that it has completed its actions to establish secure communications.

Once the client informs the server that its actions to establish secure communications are complete, the server needs to respond in a similar manner. To do so the server transmits a Change Cipher Sec message to inform the client to begin using the previously agreed upon security methods. This is followed by the Finish message, which informs the client that the server has completed its negotiations and secure communications are in effect.

Of the nine messages shown in Figure 5.40, only two actually are used for authentication. The Certificate message, as previously noted, contains information the client can use to verify the identity of the server. In doing so the client will verify the certificate signature and other information contained in the message. For example, the signature represents a hash of the certificate encrypted by the issuer of the certificate through the use of its private key. By knowing the issuer's public key, it then becomes possible to verify the signature on the certificate, which in effect validates the certificate. To do so the client obtains the public key of the certificate authority that issued the key.

The second message that in effect verifies the identity of the server is the Client Key Exchange message issued by the client. When the client transmits this message, it encrypts data using the public key contained in the previously received and verified server certificate. This action means that only the party that possesses the server's private key can decrypt the encrypted message. Thus, this message indirectly verifies the identity of the server.

Now that we have an appreciation for SSL authentication, let us turn our attention to encryption.

5.3.4.2 Establishing Encrypted Communications

Similar to authenticating a server, establishing encrypted communications represents the exchange of a sequence of SSL messages. Figure 5.41 illustrates the sequence of SSL message required to establish encrypted communications between a client and server. In comparing Figure 5.40 to Figure 5.41 note that both actions commence with a Client Hello message. The key difference between the two is the use of the Certificate message to authenticate a server's identity while a Server Key Exchange message is used by the server when establishing encrypted communications. The Client Hello and Server Hello response convey proposed SSL options and the server's selection of those options. This is followed by the Server Key Exchange message in which the server transmits its public key to the client. The server then completes its initial sequence of messages with the Server Hello Done message.

Once the server completes its response to the initial Client Hello message, the client transmits session key information in the Client Key Exchange message.

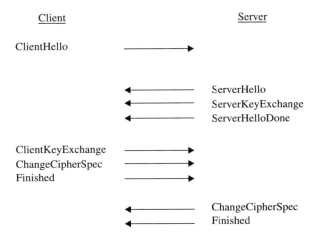

Figure 5.41   SSL messages to establish encrypted communications.

Information in this message is encrypted using the public key provided by the server. This is followed by the Change Cipher Spec message, which activates the previously negotiated options. The Finished message informs the server that the negotiation is complete. The server then responds with its Change Cipher Spec message to activate its side of the previously negotiated options. This is then followed by the server transmitting a Finished message to inform the client that the negotiation is complete.

5.3.5 Cipher Suites

As previously noted, the Client Hello message is used to propose various SSL options, with the Server Hello message selecting the options to use. Within the Client Hello message is a Cipher Suites field that denotes the cryptographic services that the client can support. Thus, the strength of SSL security can be considered to be represented by the cipher suites a client supports. Each cipher suite defines a key exchange algorithm, encryption method, and hash algorithm. SSL V3.0 defines 31 different cipher suites, which range in scope from an open system without the use of encryption to a mixture of different types of key exchanges used for different encryption methods and different hash algorithms used for supporting message authentication.

Table 5.3 lists 14 key exchange algorithms supported by SSL. The two columns in Table 5.3 indicate the algorithm and provide a brief description of the algorithm.

TABLE 5.3   SSL Key Exchange algorithms

Algorithm	Description
DHE_DSS	Ephemeral Diffie-Hellman with DSS signatures
DHE_DSS_EXPORT	Ephemeral Diffie-Hellman with DSS signatures (512 bits max key size)
DHE_RSA	Ephemeral Diffie-Hellman with RSA signatures
DHE_RSA_EXPORT	Ephemeral Diffie-Hellman with RSA signatures (512 bits max key size)
DHE_anon	Anonymous Diffie-Hellman
DHE_anon_EXPORT	Anonymous Diffie-Hellman (512 bits max key size)
DH_DSS	Diffie-Hellman with DSS certificates
DH_DSS_EXPORT	Diffie-Hellman with DSS certificates (512 bits max key size)
DH_RSA	Diffie-Hellman with RSA certificates
DH_RSA_EXPORT	Diffie-Hellman with RSA certificates (512 bits max key size)
FORTEZZA_DMS*	Fortezza/DMS
NULL	No key exchange
RSA	RSA key exchange
RSA_EXPORT	RES key exchange (512 bits max key size)

*DMS Defense Message System

The basic Diffie-Hellman key exchange method consists of a sequence of operations that enable two hosts to share a secret key. Those steps are as follows:

1. The hosts obtain Diffie-Hellman parameters in the form of a prime number p greater than 2 and a base g which represents an integer smaller than p. Both p and g can either be hard coded on a host or retrieved from a server.
2. Each host generates a private number x which is less than $p - 1$.
3. Each host generates their public key y via the following algorithm

$$y = R((g \wedge x)/P)$$

where R denotes the remainder.
4. The two hosts exchange their public keys (y), which are then converted into secret keys z using the following algorithm:

$$z = R((y \wedge x)/P)$$

The secret key (z) can be used as the key for use with a selected encryption method.

Note that an ephemeral Diffie-Hellman key exchange provides forward security. This means that the compromise of either party's key will not allow prior sessions previously secured to be decrypted. In comparison, an anonymous Diffie-Hellman key exchange does not incorporate any authentication.

As you examine the entries in Table 5.3 you will note that with the exception of two key exchange algorithms (Fortezza/DMS and Null) each algorithm has two versions. One version places no limit on the key size, while the other version of the key exchange algorithm, which is designed to be exportable, has a 512-bit key size limit. Since the US Government periodically changes its export policies with respect to encryption, readers who need to establish VPNs that include international locations should check with the Bureau of Industry and Security (BIS), a department of the US Department of Commerce. BIS publishes rules regarding export regulations of encryption products.

5.3.5.1 Encryption Algorithms

SSL Version 3.0 supports nine encryption algorithms. Those algorithms, which are listed in Table 5.4 include both block and stream ciphers.

As a refresher, a block cipher requires an initialization sector of dummy data to begin the encryption process. In addition, a block cipher is restricted to operations on fixed length blocks of data. In comparison, a stream cipher represents the process whereby each byte of input is operated upon, enabling this method of encryption to accept any length of data for processing. DES, 3DES, and RC2 represent block ciphers while RC4 represents an example of a stream cipher. Because the advanced Encryption Standard (AES) was only recently standardized and its complexity may require the use of a co-processor it is not currently supported.

TABLE 5.4 SSL encryption algorithms

Algorithm	Type	Key size (bits)	IV size (bytes)
3DES_EDE_CBC	Block	168	8
DES_CBC	Block	56	8
DES40_CBC	Block	40	8
FORTEZZA_CBC	Block	96	20
IDEA_CBC	Block	128	8
NULL	Stream	0	N/A
RC2_CBC_40	Block	40	8
RC4_128	Stream	128	N/A
RE4_40	Stream	40	N/A

5.3.5.2 The Hash Algorithm

The third and final component of an SSL cipher suite is the hash algorithm. The hash algorithm is provided for message authentication. As a refresher, a hash algorithm operates upon a variable length string to produce a fixed length number. That number represents the hash of the input string such that a small change in the input results in a substantial change in the hash value. SSL supports MD5, SHA and a NULL method. MD5 has a 16-byte hash, while SHA uses a 20-byte hash.

Now that we are familiar with the three components of an SSL cipher suite, let us turn our attention to the 31 defined cipher suites. Those suites are listed in Table 5.5.

Fortezza in Italian is the term for 'fortress'. Fortezza represents a family of security products trademarked by the US Government's National Security Agency (NSA) to include crypto cards that encrypt all files on a PC as well as a block encryption algorithm that several vendors now offer in software. DES represents the original 56-bit Data Encryption Standard that can now be successfully attached in a brute force trial and error method, while 3DES represents an extension to DES that in effect triples the key length. Thus, SSL cipher suites that support 3DES encryption would be preferable to a suite that supports DES. Similarly, because MD5 creates a 16-byte hash while SHA produces a 20-byte hash, therefore the latter provides a higher level of security.

Now that we have an appreciation of SSL let us turn our attention to the use of an SSL compatible network appliance. Specifically, we will examine the operation and potential utilization of the Netilla Security Platform.

5.3.6 The Netilla Security Platform

The Netilla Security Platform (NS) represents a popular SSL VPN network appliance. This network device is referred to as a clientless SSL VPN appliance as it enables any person using an SSL compatible browser to establish a secure connection to the device. Thus, the use of the appliance does not require any special software to be installed on a client.

5.3.6.1 Operation

The Netilla Service Platform is designed for installation behind the router or gateway that provides access to a corporate network. As illustrated in Figure 5.42, the NSP connects to both the WAN and the LAN. This connection can occur using either a common Ethernet port or two separate Ethernet ports.

TABLE 5.5   SSL Cipher suites

Key exchange	Encryption	Hash
SSL_NULL_	WITH_NULL_	NULL
SSL_RSA	WITH_NULL_	MD5
SSL_RSA	WITH_NULL_	SHA
SSL_RSA_EXPORT	WITH_RC4_40_	MD5
SSL_RSA	WITH_RC4_128_	MD5
SSL_RSA	WITH_RC4_128_	SHA
SSL_RSA_EXPORT	WITH_RC2_CBC_40	MD5
SSL_RSA	WITH_IDEA_CBC	SHA
SSL_RSA_EXPORT	WITH_DES40_CBC	SHA
SSL_RSA	WITH_DES_CBC	SHA
SSL_RSA	WITH_3DES_CBC	SHA
SSL_DH_DSS_EXPORT	WITH_DES40_CBC	SHA
SSL_DH_DSS_	WITH_DES_CBC	SHA
SSL_DH_DSS_	WITH_3DES_EDE_CBC_	SHA
SSL_DH_RSA_EXPORT	WITH_DES40_CBC_	SHA
SSL_DH_RSA	WITH_DES_CBC	SHA
SSL_DH_RSA	WITH_3DES_EDE_CBC	SHA
SSL_DHE_DSS_EXPORT	WITH_DES40_CBC	SHA
SSL_DHE_DSS_	WITH_DES_CBC_	SHA
SSL_DHE_DSS_	WITH_3DES_EDE_CBC	SHA
SSL_DHE_RSA_EXPORT	WITH_DES40_CBC_	SHA
SSL_DHE_RSA_	WITH_DES_CBC_	SHA
SSL_DHE_RSA_	WITH_3DES_EDE_CBC_	SHA
SSL_DH_anon_EXPORT	WITH_RC4_40	MD5
SSL_DH_anon_	WITH_RC4_128_	MD5
SSL_DH_EXPORT	WITH_DES40_CBC_	SHA
SSL_anon_	WITH_DES_CBC	SHA
SSL_anon_	WITH_3DES_EDE_CBC_	SHA
SSL_FORTEZZA_DMS_	WITH_NULL_	SHA
SSL_FORTEZZA_DMS_	WITH_FORTEZZA_CBC_	SHA
SSL_FORTEZZA_DMS_	WITH_RC4_128_	SHA

Remote clients communicate with the NSP by entering the host name or IP address of the appliance into their browser to set up an SSL encrypted session. The NSP functions as a proxy, terminating the SSL session and communicating directly to applicable application servers on behalf of the client.

In the example shown in Figure 5.42 the remote LAN client, which could represent an employee at a branch office, is using the Netilla Service Platform via an Internet connection to access an application server at a corporate central site. In comparison, a dial modem user that could represent an

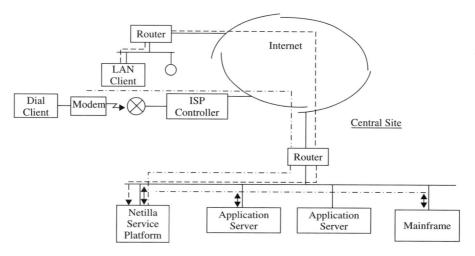

Figure 5.42 Using the Netilla Service Platform.

employee anywhere on the globe is shown connecting through an Internet Service Provider (ISP) to the NSP at the corporate central site to access a mainframe. For both clients, communications from each location to the NSP at the central site occurs via SSL, providing authentication and encryption, which secures communications via the Internet.

The Netilla Service Platform supports communications with a wide range of application servers. Some of those applications include 3270 emulation for mainframe access, UNIX/Linux, X Windows and character-based application support and the ability to access Windows 2000 Terminal Services. The Netilla Service Platform can be accessed via any Internet connection method, such as a cable modem or DSL connection or from a client connected to the Internet via a LAN.

5.3.6.2 Authentication

Prior to a client gaining access to the NSP they must be authenticated. Authentication occurs through the use of a user-name and associated password, with the user population subdivided into two categories: local and external. Local users have accounts that are created and stored on the NSP. Therefore, authentication occurs on that device. In comparison, external users represent clients for whom accounts are established in a database external to the NSP, such as a Windows 2000 domain controller, or a Kerberos or Secure ID server. Regardless of the client authentication category, all users will have a profile that contains configuration information that defines the client display upon their connection to the NSP. Thus, the administrator can tailor the configuration of each client to control access to all or a subset of the applications supported by this network appliance. Typically, for the support of a large number of clients, the administrator would create a series of groups, associate one or more applications to each group and then associate a group membership to each user.

5.3.6.3 Applications

As previously mentioned, the Netilla Service Platform supports a variety of applications. In this section we will describe some of those applications in more detail, illustrating how the use of this network appliance provides secure access to applications that otherwise could be difficult to secure.

X-Windows

The NSP can provide access to a range of applications installed on an X-Windows platform that uses the X11 Windowing Protocol. Examples of applications that can be supported include Telnet, rexec, ssh, rcmd and rlogin. Because these applications on their own do not support security they could represent a danger to an organization when used via the Internet. For example, consider a remote login (rlogin) session in which a client directly accesses a host. Because data flows over the Internet, interception is possible whose compromise could allow a third party to obtain unauthorized access to an organizational host. However, when the NSP is used, data between the client and the NSP is encrypted, with clear-text only flowing on the organization's internal network, and this action eliminates external threats.

3270 Terminal emulation

Another service offered by the NSP is support for the 3270 protocol. In addition to providing a proxy service for applications on a mainframe computer that supports 3270 terminals, the 3270 protocol permits applications on IBM AS/400 minicomputers to be accessed. Although the IBM AS/400 uses the 5250 protocol, that computer series also supports access via a 3270 terminal that emulates the 5250. By defining the appropriate mapping of terminal keys, a process referred to as keymap, remote clients can access most of the features used in a 5250 terminal access session.

UNIX

Currently the Netilla Service Platform is limited to supporting UNIX character-based applications. Such applications can include Telnet, ssh, rexec, rcmd and rlogin.

5.3.7 Summary

Because the NSP supports SSL, there is no need for clients to be configured with specialized software, such as IPSec. This in itself can significantly reduce hardware and software requirements since most organizational PCs can support modern browsers without modification. Since the software and hardware needed to set up a VPN can be much more difficult than launching a browser, the NSP and similar network appliances reduce potential remote client support to a simple browser address entry. At the central site, the ability to access multiple applications as well as the ability to manage user access by groups simplify the management process. Since all remote client-to-platform communications occur using 128-bit encryption built into modern browsers and use digital certificates for authentication in a transparent manner, VPN client support efforts are eliminated. At the central site, the steps required to configure the NSP occur through the use of a series of menus that are similar in scope to the configuration process associated with other networking devices. While the network manager or LAN administrator now has another device that they are responsible for, the ability to provide a VPN capability to numerous applications from both mobile and fixed location clients without the necessity to configure and maintain client software can usually result in an overall level of required support significantly less than alternative VPN methods.

chapter six

VPN Hardware and Software

The purpose of this chapter is to become acquainted with representative hardware and software products that can be used to create different types of VPNs. Because there are literally hundreds of products available for consideration, the aim of examining hardware and software products is to show readers different methods you can consider for creating VPNs instead of providing a feature comparison of products. For example, in the first section in this chapter we will describe and discuss the Asanté Friendly Net VR2004 Series VPN Security Routers, which can be obtained with four 10/100 LAN ports plus integrated IEEE 802.11b wireless access point to provide secure communications for wired and wireless devices. In reality, as we will shortly note, the VR2004 router establishes a secure connection to another device that provides a compatible VPN solution. Thus, security from a client behind the router communicating via radio frequency (RF) to the router can be considered to represent a potential problem. In the remainder of this chapter we will briefly examine the operation and use of a remote access VPN solution available with the Windows operating system. Thus, this chapter will provide readers with a foundation concerning VPN options that can support the networking requirements of their organizations.

6.1 Using the ASANTE VPN Security Router

There are many types of routers that include a built-in VPN capability. Most routers that offer this capability require the user to issue a series of commands to configure the device. What is most interesting about the router we will examine in this section is the fact that it uses menus to provide the user with the ability to establish an appropriate VPN. In addition, as noted during the introduction to this chapter, the router we will examine in this section

Virtual Private Networking   G. H. Held
© 2004 John Wiley & Sons, Ltd   ISBN: 0-470-85432-4

includes IEEE 802.11b wireless LAN support, permitting both wired and wireless clients at one geographical location to obtain a secure communications capability across the Internet to another geographical location.

The basic functionality of the Asanté Friendly Net VR2004 Series VPN Security Router is similar to other wireless access points used by this author when he was researching a book he was writing on wireless LAN security. That is, the router is pre-configured with an RFC 1918 Class C address, supports DHCP via a built-in DHCP server capability which permits IP addresses to be automatically assigned to clients and includes a network address translation (NAT) capability. What sets this router apart from other wireless LAN routers is its built-in VPN capability.

6.1.1 Overview

The Asanté VPN Security Router is available in two configurations. The first configuration includes four 10/100 Ethernet LAN ports and a backup modem port built into the router housing which is approximately the size of two packs of cigarettes. The second router configuration adds an integrated IEEE 802.11b wireless access point and dual antennas to the common router housing. Because both configurations include built-in IPSec support, which is the focus of this book, our discussion of both routers will be primarily on this feature. However, because wired and wireless clients have different vulnerabilities we will discuss both with respect to client-to-router transmission later in this section.

6.1.2 Configuration access

Both versions of the Asanté VPN Security Router can be configured via the use of a Web browser. Because both routers are configured by default with the RFC 1918 IP address of 192.168.123.254, you would enter that address into the browser's address box to access the router. This action will result in the display of a login window that prompts you to enter a password.

Asanté Technologies is similar to other hardware vendors in that their equipment is shipped with some pre-configured default settings. In particular, the default password is set to 'admin' and should obviously be changed during the initial router configuration process. Once you enter the applicable password the router's main menu will be displayed.

Figure 6.1 illustrates an example of the Asanté VPN Security Router's main menu for the vendor's product that includes support for IEEE 802.11b clients.

The Asanté VPN Security Router that is limited to supporting wired clients has a similar main menu but lacks support for wireless settings. Not shown

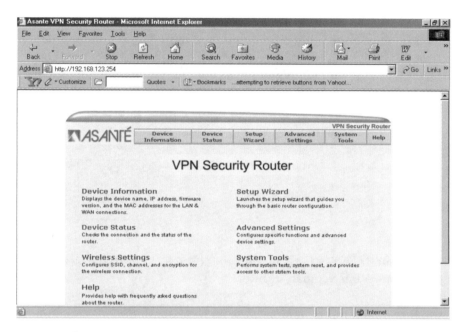

Figure 6.1 The Asanté VPN Security Router main menu.

in Figure 6.1 are two entries at the bottom of the menu that are hidden from view. One entry is labeled 'Launch the Setup Wizard' and duplicates the entry at the top of the second column. The selection of either entry invokes a setup wizard. That wizard can be used to configure time zone settings, device IP settings, static IP settings, VPN settings and other settings. The second entry that is hidden from view is labeled 'Logout' and its selection logs you out of the router and terminates your ability to configure the device.

6.1.3 Wireless considerations

In a wireless environment remote clients communicate with the router via RF connections. The router can then be configured to communicate with a distant location using IPSec to secure communications through the Internet. Figure 6.2 illustrates the use of the Asanté VPN Security Router with built-in IEEE 802.11b support, indicating the relationship between wireless security and IPSec provided by the router.

In examining Figure 6.2 note that wireless LAN security functions supported by the router protect communications occurring over the air from the client to the router. In contrast, security for transmission from the router through the

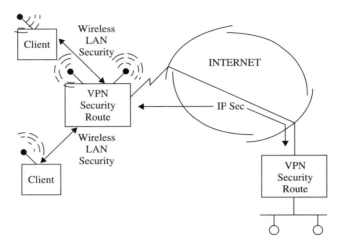

Figure 6.2 Relationship between wireless LAN and IPSec security when using the Asanté VPN Security Router.

Internet is provided by its IPSec functionality. Thus, to obtain an appreciation of end-to-end security we will examine both as well as digress a bit and note why in some situations the use of IPSec on an end-to-end basis may be more appropriate.

6.1.3.1 Wireless Settings

Figure 6.3 illustrates the Asanté VPN Security Router wireless settings menu, which is applicable for the router that includes IEEE 802.11b support.

The SSID
By default, the router uses the text 'Wireless' for its service set identifier (SSID). Because an access point is typically limited to serving between 20 and 30 clients, organizations commonly install multiple access points within a limited geographic area in a building to service a larger client base, with each access point configured with a unique SSID. Thus, the SSID permits a client to access or become associated with a particular access point. Since the client must be configured with the SSID of a particular access point, some literature refers to the SSID as a password. Unfortunately, it is a rather poor password since it is transmitted in the clear. In addition, the SSID is usually set to a default value that is documented on vendor Websites and can be easily overwritten by configuring the client to either a blank SSID or using 'any' as

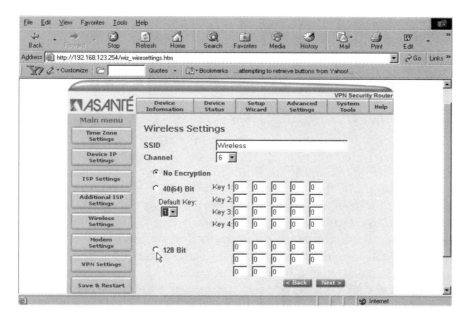

Figure 6.3 Asanté's Security Router wireless settings involve configuring an SSID and selecting either a static 64-bit or 128-bit encryption key.

the SSID. The use of a blank SSID results in the display of the SSIDs of other access points within RF range of the client, permitting the client to select an appropriate AP. Thus, there are several methods an unauthorized third party can employ to determine the SSID in use or even override its use.

Encryption
Returning our attention to Figure 6.3, note that by default the Asanté VPN Security Router disables encryption. For readers familiar with wireless LAN equipment this should come as no surprise, since most vendors disable encryption, requiring customers to enable this feature. Although the Asanté VPN Security Router supports both 64-bit and 128-bit encryption referred to as wired equivalent privacy (WEP), the use of both encryption methods are extremely vulnerable to compromise. In fact, several programs are available via the Internet that can be used to recover a previously configured WEP key via passive monitoring of RF signals. Each of these programs requires between 4 and 5 million frames to be captured, forming a database for analysis that results in the recovery of the WEP key in use. Once the key is recovered, an unauthorized third party can 'read' the contents of previously recorded

frames as well as the contents of future transmissions that use the same encryption key.

To improve WEP several vendors now rotate encryption keys, in effect permitting their use for a limited number of frames. While the IEEE is working on a new method to increase wireless LAN security in the form of the 802.11I standard, several vendors now provide an interim solution referred to as the Temporal Key Integrity Protocol (TKIP). Although a discussion of TKIP is beyond the focus of a book on VPNs, it should be mentioned that this standard involves the periodic change of WEP keys. Thus, organizations that are concerned about both the potential monitoring of their clients' transmissions within a building and the security of transmission over the Internet should probably consider creating a VPN from the client or using a wireless access point that supports a more robust and secure use of WEP. However, to be fair to Asanté Technologies, it should be noted that their VPN Security Routers include two additional features that can be used to control network access and in effect provide a degree of additional network security. One feature, referred to as 'wireless access control settings' permits the network manager or LAN administrator to either enable or disable access to the wireless network based upon pre-defined MAC addresses entered into the router. Although you can use this feature to represent MAC address authentication, it is important to note that under the IEEE 802.11b standard MAC addresses are transmitted in the clear and can be easily monitored. Thus, while wireless access control settings makes it more difficult for an unauthorized third party to gain access to a network, it does not prevent them from doing so. A more robust authentication method would be equipment that supports the IEEE 802.1x standard and the use of a RADIUS server. While other vendor access point products support the 802.1x standard, to this author's knowledge, they do not incorporate IPSec support. Thus, like life, equipment acquisition represents a trade-off.

A second feature included in the Asanté VPN Security Router that deserves mentioning is its filtering capability. You can establish LAN filters to enable or disable certain types of access to the Internet or restrict traffic from the Internet. For either situation you can establish a filter by specifying three metrics. Those metrics are the protocol to be filtered (IP, TCP, or UDP), an IP address range and destination port range. By assigning static IP addresses to LAN clients and using LAN filtering and the previously described wireless access control settings for MAC authentication, you can make it more difficult for an unauthorized third party to gain access to your organization's network.

Now that we have an appreciation of a few wireless LAN security problems, let us turn our attention to the VPN capability of the Asanté router.

6.1.4 VPN operations

The Asanté VPN Security Router supports both network-to-network, also referred to as site-to-site and client-to-network VPN creation. As a refresher, the latter permits clients operating IPSec software to create an IPSec VPN tunnel to a central site IPSec VPN compliant device, such as an Asanté VPN Security Router. In this section we will discuss the use of the Asanté router and its configuration with respect to both VPN networking modes of operation.

6.1.4.1 Network-to-Network

In a network-to-network operating environment two Asanté VPN Security Routers are configured to work with one another to create a tunnel through the Internet. To facilitate providing an example of the configuration of a pair of Asanté routers let us assume our organization has two geographically separated locations that will be interconnected over the Internet. For each location you will need three items of information to configure each router. Those items of information are:

1. The remote IP network address of each LAN. This is usually the address of the LAN port connected to the router, with the last dotted decimal position changed to '0'.
2. The remote LANs subnet mask.
3. The remote routers' WAN IP address.

6.1.4.2 Need for Address Modification

Let us assume the two networks to be interconnected are located in Macon, GA, and Chicago, IL, and have LAN and WAN addresses as noted in Figure 6.4.

If you carefully examine Figure 6.4 you will note that both LANs are shown assigned the same RFC 1918 Class C address, which is the default address used by Asanté routers. If you remember our prior examination of the use of IPSec in this book, the creation of an IPSec tunnel results in a new header prefixed onto the IP packet. Thus, the addresses in the IP header flow from end-to-end unmodified by any network address translation performed by the local router. If we do not modify at least one LAN default address, it would then appear that both networks have the same address, resulting in the inability of routers to route packets off each network. Because each network joined by VPNs must have a different network address, one of the two LAN addresses must be changed. The Asanté router is similar to several other vendor products examined by this author in that it supports all three blocks of RFC 1918 addresses. Those addresses are listed in Table 6.1.

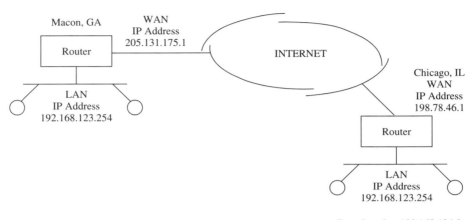

Figure 6.4 When configuring routers you cannot keep the LAN IP addresses at each end set to the same default IP address.

TABLE 6.1 RFC 1918 addresses

10.0.0.0 to 10.255.255.255
172.16.0.0 to 172.31.255.255
192.168.0.0 to 192.168.255.255

To facilitate changing the LAN address on one router connection you can simply change the third dotted decimal in one of the default address assignments. For illustrative purposes we will assume the LAN in Macon is assigned the default network address of 192.168.123.0 while the Chicago LAN's network address was changed to 192.168.124.0. Because each LAN address represents a Class C address, the subnet mask is 255.255.255.0 for each network. From Figure 6.4 we will assume that the WAN IP address of one router is 205.131.175.1 while the WAN IP address of the second router is 198.78.46.1.

6.1.4.3 VPN Settings

The initial VPN Settings screen on an Asanté router is shown in Figure 6.5. Note that after you enter a connection name and click on the button labeled 'Add' the display will change to the one similar to Figure 6.6. However, prior to discussing the VPN settings fields shown in Figure 6.5 a few words are in order concerning the lower portion of Figure 6.5.

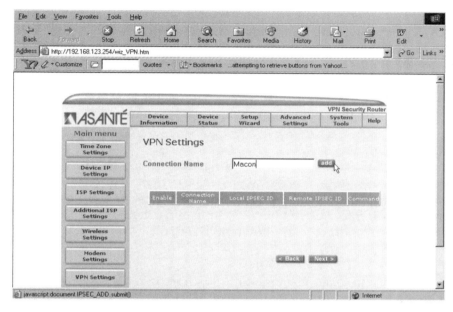

Figure 6.5   The Asanté Security Router's initial VPN Settings display.

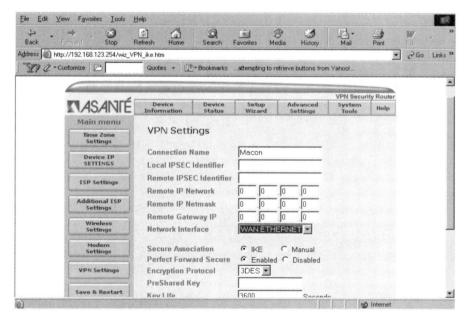

Figure 6.6   The Asanté Router VPN Settings configuration screen.

If you look at the horizontal bar you will note five labels, beginning with 'Enable'. Once you configure a VPN connection, information concerning the connection will be displayed below the highlighted bar with the connection name used as a reference for a particular predefined VPN.

Now that we have an appreciation of the initial screen, let us turn our attention to the second VPN Settings screen which provides us with the ability to configure the connection. That screen is shown in Figure 6.6.

Connection name

In examining the fields in the VPN Settings display shown in Figure 6.6, the connection name can be considered as an identifier. This name provides a reference that enables a router administrator to have several pre-defined VPNs and enable or disable a particular VPN from a list that will appear under the bar labeled 'Enable' previously shown in Figure 6.5.

Local and remote IPSec identifiers

The purpose of the local and remote IPSec identifiers is to provide a mechanism to identify multiple tunnels. Neither identifier needs to match the name used at the other end of the tunnel. The default value for the local IPSec identifier is 'Local,' while the default for the remote IPSec identifier is 'Remote.'

Network addresses

Assuming we are configuring the Macon, GA, router located on the left side of Figure 6.4, the remote IP network address would appear to be 192.168.123.254. However, as previously explained, the Asanté VPN Security Router sets each connected LAN to the IP network address of 192.168.123.0. Because you cannot have two networks with the same IP address, you would need to configure one router to use a different DHCP IP network address. If we assume the router located in Chicago, IL, and which is shown located on the right portion of Figure 6.4 assigns addresses to LAN clients using the IP address of 192.168.100.0, then you would enter that address in the remote IP network field. Since that address is a Class C address, you would enter the remote IP netmask as 255.255.255.0. The remote gateway IP address would then be entered as 198.78.46.1, since that is the IP address assigned to the remote router as per our example shown in Figure 6.4.

Network interface

The network interface field shown in Figure 6.6 is actually a pull-down menu. Because the Asanté router is designed to support cable and DSL modem connections, you would use its default network interface setting of 'WAN ETHERNET' as shown in Figure 6.6.

Once you configure the entries located at the upper portion of the VPN Settings display, you would use the lower portion of the display for configuring IPSec settings.

IPSec settings

The first entry is labeled 'Secure Association' and involves the Security Association setting. This setting can be set to either IPSec Keying (IKE) or manual mode via a radio button. If you select the default setting of IKE, the key used for encryption when a secure association is formed in each direction will be automatically distributed between routers. However, because some ISPs block TCP port 500 traffic on their firewalls, it might not be possible to transmit keys via IKE. You would then select the manual radio button. Concerning the latter, when you set the secure association to manual you then need to specify both an incoming and outgoing Security Parameter Index (SPI) as 8-hex digit numbers. The incoming SPI must match the outgoing SPI value at the other end of the tunnel. Similarly, the outgoing SPI value must match the distant end incoming SPI value. During the manual setup process you also need to specify the encryption protocol, encryption key, authentication protocol and authentication key.

The encryption protocols supported by Asanté routers include 3DES (triple DES), DES and Null, the latter representing no encryption. When manually setting up a key exchange you would enter 24 alphanumeric characters for a 3DES setting or 8 characters for a DES setting as the encryption key for both routers with each end is obviously configured with the same setting. You would then use another drop-down menu to select one of two authentication algorithms supported by Asanté routers – MD5 or SHA-1. The selected authentication protocol must obviously be the same on both routers. The last metric that needs to be specified for a manual mode secure association is an authentication key. If you selected MD5, you would enter 16 alphanumeric characters, while the selection of SHA-1 would require the entry of 20 characters. Once again, both routers would be configured with the same value.

Returning our attention to Figure 6.6, assuming you selected IKE for the secure association, let us examine the other entries in the display. The Perfect Forward Secure (PFS) represents an optional feature of IKE that, when enabled, provides added protection against an unauthorized third party monitoring and attempting to decode encrypted data. When enabled, which is the default setting, it must match the remote router's setting.

The encryption protocol pull-down menu is the same as previously discussed for a manual secure association. That is, you can set the protocol to null, 3DES or DES. The pre-shared key field permits you to enter an initial key that is used for both encryption and authentication. That key can be up to 256 characters and its value must match the entry on the remote router. The following field, Key Life, sets the amount of time until the router re-negotiates the key. The default value is 3600 seconds or one hour. Obscured from view in Figure 6.6 is the last VPN setting you need to specify. That setting concerns IKE Life time, which is used to specify the amount of time until the router re-negotiates the IKE security association. The default value of this field is 28800 seconds or 8 hours.

Configuration summary

Assuming we are connecting LAN locations in Macon and Chicago as previously shown in Figure 6.4, we can summarize our router configurations. Table 6.2 provides a summary of configuration settings for each of the routers shown in Figure 6.4, assuming that we changed the network IP address of the LAN in Chicago to 192.168.124.0 and accepted each router's default settings and a pre-shared key set to abadabado2you. While the pre-shared key appears to be familiar, note that it includes a numeric and a suffix to the common prefix. The purpose of adding one or more numerics to an IPSec pre-shared key is to make it more difficult for an unauthorized third party monitoring communications to attempt to decipher captured data.

TABLE 6.2   VPN Security Router settings

Setting	Macon router	Chicago router
Connection name	Macon	Chicago
Local IPSec Identifier	Local	Local
Remote IPSec Identifier	Remote	Remote
Remote IP Network	192.168.123.0	192.168.124.0
Remote IP Netmask	255.255.255.0	255.255.255.0
Remote Gateway IP	205.131.175.1	197.78.46.1
Network Interface	WAN ETHERNET	WAN ETHERNET
Secure Association	IKE	IKE
Perfect Forward Secure	Enabled	Enabled
Encryption Protocol	3DES	3DES
Pre-shared Key	abadabado2you	abadabado2you
Key Life	3600	3600
IKE Life Time	28800	28800

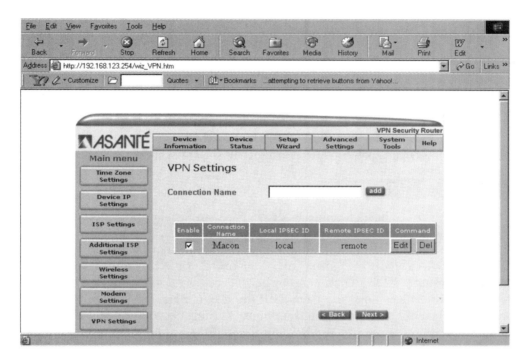

Figure 6.7 Upon completing a VPN configuration the configuration can be edited or deleted from the VPN Settings screen.

Once you complete the configuration of VPN settings for a particular router, the display will change to provide a summary of the previously established configuration. An example of this summary is shown in Figure 6.7, which now indicates that the connection we named 'Macon' on the router at that location is enabled. Note that you can easily edit or delete the configuration by clicking on an appropriate button.

6.1.5 Client-to-network

The client-to-network access method can be considered to represent a VPN remote access method. In a client-to-network IPSec mode the configuration of the Asanté router used to terminate the client will depend upon whether IP client address assignments occur dynamically via DHCP or clients have static (fixed) IP addresses.

6.1.5.1 Dynamic Client IP Address

If the remote client wants to obtain a dynamic IP address, you would configure each of the router's remote IP network, remote IP netmask and remote gateway IP addresses to 0.0.0.0. This action is necessary since there is no way to pre-define the IP address of a remote client. You would then configure the network interface, local IPSec and remote IPSec identifiers in the same manner as previously described for the network-to-network configuration. Of course, the remote client must obtain and configure IPSec software to access the router.

6.1.5.2 Static Client IP Address

If the remote client has a fixed IP address, the user must configure IPSec client software that functions as a virtual NIC card to access the remote Asanté router. That is, the software must enable the remote client to appear to the router as a virtual NIC card. In this situation you would configure the remote router so that its remote IP network, remote netmask and remote gateway IP entries reflect the IP addresses associated with the client's virtual LAN and static IP addresses. For example, assume the client PC had the virtual LAN IP address of 192.168.123.0 while the WAN IP address of the router serving the client's network was 205.131.175.5, which is also a Class C network address. Under this scenario you would configure the routers' remote IP network address as 192.168.123.0, which is the client's virtual LAN IP address. The routers' remote netmask would be set to 255.255.255.0, while the remote gateway IP address would be configured to 205.131.175.0. The other router parameters, such as its network interface and local and remote IPSec identifiers, would be configured as previously noted.

Now that we have an appreciation of the operation and configuration of the Asanté VPN Security Router, let us turn our attention to the creation of VPNs under different versions of the Windows operating system.

6.2 Windows VPN software

Over the years Microsoft has included VPN capabilities bundled with its various PC client and server platforms. On the client side, a VPN capability was built into Windows 98 and was enhanced under Windows 2000, Windows ME, and Windows XP. On the server side, Microsoft included a VPN capability in Windows NT, Windows 2000 and its recently introduced Windows Server 2003 product. Concerning the latter, improvements in Windows Server 2003 VPN software include an expansion of the ability to move VPN traffic through firewalls, support of additional authentication methods and the denial of access to the server VPN side when the client attempting to connect is not configured with appropriate security applications, such as a firewall and anti-virus software. Regardless of the VPN feature set, each Microsoft server functions in a similar manner, as a VPN gateway that terminates client sessions. The key difference between server VPN capabilities resides in the authentication methods and security protocols supported. Because Microsoft is expected to release both client and server upgrades that will provide a common set of features for people using different client products, this author decided to primarily focus attention upon the use of Windows XP as a client platform for accessing a server in this section. Concerning the server, due to the popularity of Windows 2000 this author will illustrate its use in functioning as a termination point for remote clients.

6.2.1 Using a Windows XP client

At the time this book was prepared, Windows XP represented the latest version of Microsoft client software. In this section we will turn our attention to the use of a Windows XP client to create a VPN to a Windows 2000 server. In doing so this author will use a Cisco wireless LAN adapter in his laptop to illustrate how a VPN can secure communications occurring at WiFi 'hotspots' located in coffee shops, airports and hotels.

6.2.2 Creating the VPN

To create a VPN connection under Windows XP you would click on Start, point to Settings; click on Control Panel, and then double-click on Network Connections. From Network Connections you would select 'Create a new connection' which is located in the Network Tasks column on the left portion of the screen. This action invokes a New Connection Wizard, which is shown in the center of Figure 6.8.

In examining Figure 6.8 note that the background portion of the screen to the left of the New Connection Wizard box indicates this author is using a wireless network connection which was enabled when the screen was captured. The wireless network connection occurred using a Cisco Systems 350 series wireless LAN adapter that communicated with an access point located in this author's home. However, a similar process is applicable for use in hotels, airports and your favorite coffee shop hot spot. Thus, once the VPN is set up, this author can simply click on an icon to establish a secure VPN tunnel to his server from hotels, airports and his favorite coffee shot as he travels.

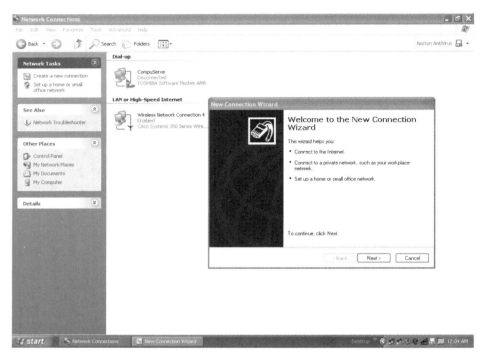

Figure 6.8   Creating a new communications connection under Windows XP is simplified via the use of a wizard.

6.2.2.1 Using the New Connection Wizard

Although the use of a wizard by its name is designed to simplify the configuration process, as we will shortly note, you need to consider several configuration parameters when you set up your VPN connection.

Specifying a Connection Type

After you click on the Next button shown in the New Connection Wizard illustrated in Figure 6.8, you will be prompted to enter the type of connection you want to create. This new display, which is titled 'Network Connection Type,' is shown in Figure 6.9. Note that the second radio button is shown selected since we want to establish a VPN connection.

Other types of communications supported by the Network Connection Type dialog box include a standard Internet connection which is associated with the top radio button, setting up a home or small office network which is associated with the third radio button, and connecting your computer directly

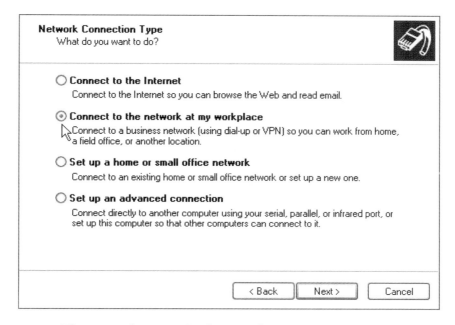

Figure 6.9 The second screen in the Windows XP New Connection Wizard prompts you to specify the type of network connection to be established.

to another via a serial, parallel or infrared port which is associated with the fourth radio button.

Specifying the communications facility

Once you select the radio button associated with the label 'Connect to the network at my workplace,' you will need to specify the communications facility you will use to establish the connection.

Figure 6.10 illustrates the Network Connection dialog box generated by the wizard. Even though we previously selected the radio button in Figure 6.9 associated with 'connecting to the network at my workplace' we need to define how that connection will occur.

In Figure 6.10 we have two options to consider – establishing a dial-up connection or connecting via a VPN connection. Since this author was using his laptop computer via a wireless LAN connection to communicate with an access point that was connected to a cable modem which in turn was connected to the Internet, the second radio button was selected. Otherwise, if this author intended to directly dial the server via a modem or ISDN connection, the top radio button shown in Figure 6.10 would have been selected.

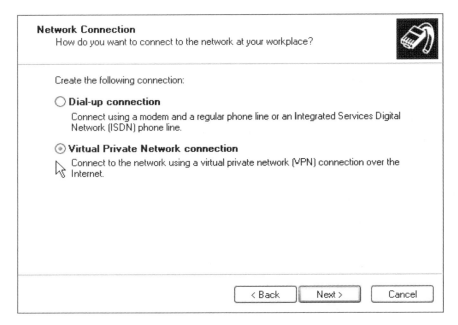

Figure 6.10   The network connection dialog box provides you with a mechanism to specify how you intend to connect to the network at your workplace.

One of the more interesting problems some readers may face is the fact that they use the same laptop from both home and when traveling to access the corporate network. To facilitate operating under this situation you should consider configuring two or more VPN connections, each appropriately labeled. Then, you could simply select one of the pre-defined connection types based upon the location from which you intend to establish a VPN connection. In reality, as we will shortly note in this section, you can also consider modifying a pre-defined VPN connection that corresponds to a generalized VPN access method for the corporate network. To accomplish the previously mentioned task you can assign an applicable name to a VPN connection and later modify the properties of the connection. You can associate a name to a VPN connection by using the next dialog box displayed by the wizard, which is labeled 'Connection Name.'

Specifying a connection name

Figure 6.11 illustrates the Connection Name dialog box. In this example this author is shown entering '4-Degree Consulting' as the workplace connection name. If this author intended to use the same laptop for traveling, he

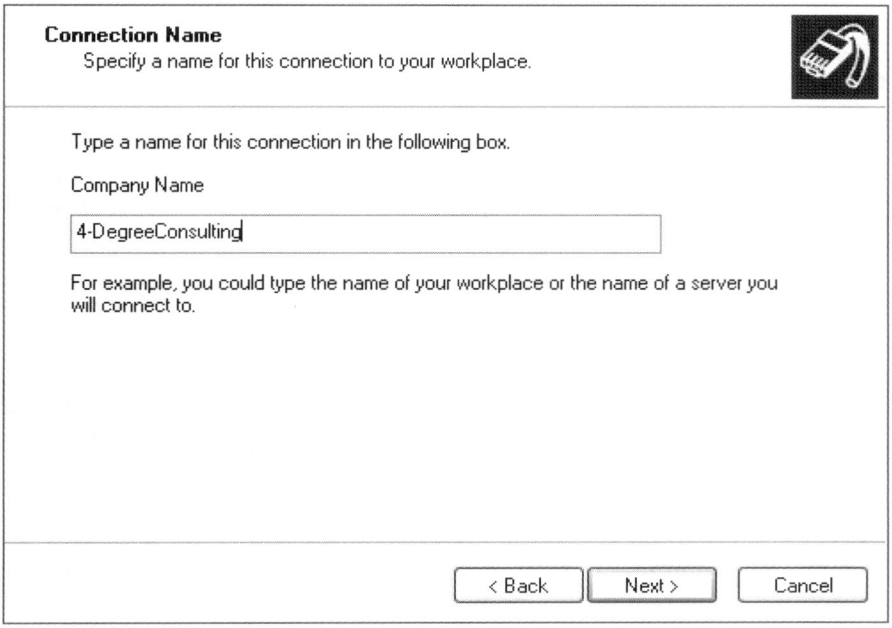

Figure 6.11   By specifying an appropriate connection name you can establish multiple, pre-configured VPNs that can be easily selected to satisfy a particular operational requirement.

would probably add the suffix 'via home' to the connection name shown in Figure 6.11.

Then, a second VPN created to support dial-up connections when on the road might have the suffix 'dial' added to a new configuration created to access the same workplace. As an alternative, if this author needed to access several servers from multiple locations, he might configure a few generalized VPN configurations, one for each server. Then, he could modify each configuration via the use of the properties of the configuration to reflect his method of access or another feature requiring alternation.

Once you specify a connection name, the wizard will display a dialog box labeled Public Network. This dialog box provides you with the ability to specify if Windows should automatically dial the initial connection to the Internet or another network prior to establishing the virtual connection. Because this author is establishing a VPN over a cable modem connection to the Internet, he selected the radio button associated with the 'do not dial the initial connection' entry. If you select the radio button associated with the 'automatically dial this initial connection' entry you would then need to specify the telephone number to be dialed.

Specifying a destination

After you click on the Public Network display's next button, the Wizard will generate a screen labeled VPN Server Selection. The purpose of this screen is to provide you with the ability to specify the destination computer you wish to access at the other end of the VPN tunnel. An example of the VPN Server Selection screen is shown in Figure 6.12.

In this example the IP address of the VPN server is shown entered, however, as indicated in the display wording, you can alternatively enter the computer's host name. If you need to access multiple computers you can use the wizard to create several configurations and assign applicable names to each configuration. Thus, the ability to assign connection names provides you with the opportunity to tailor several pre-defined configurations to your operational requirements.

As an alternative to the preceding, you can create a 'general' VPN connection that reflects the configuration required to securely access several servers. Then,

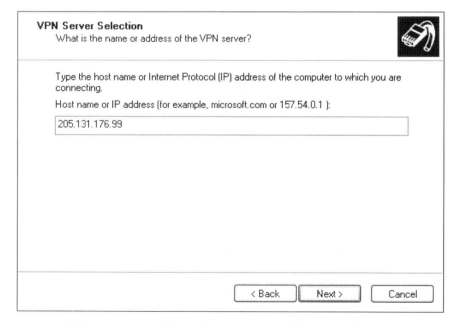

Figure 6.12   You can specify either an IP address or host name for the VPN server.

you can modify the Properties associated with the 'general' configuration to change the destination server to reflect the VPN connection you wish to establish at a particular point in time. Later in this section after we have completed our initial use of the wizard and saved our configuration, we will turn our attention to modifying its properties.

The Finish that may not be

Returning our attention to the VPN Server Selection screen shown in Figure 6.12, after you enter an applicable server address and click on the next button, the wizard will display a completion screen. This screen, which is illustrated in Figure 6.13 indicates the name you previously assigned to the connection as well as provides you with the ability to add a shortcut to the connection on your desktop. Although the screen indicates that clicking on the button completes the connection, it is important to note that your configuration uses default settings that may not be appropriate for your application. Thus, once you click on the button labeled Finish, you will more than likely wish to consider examining the properties associated with the previously created configuration and adjust those properties to your VPN environment.

Figure 6.13   When completing the new connection, you can add a shortcut to your desktop.

6.2.2.2 Examining VPN Properties

After you click on the button labeled Finish, the previously configured connection will be shown in the Network Connections display. Figure 6.14 illustrates both the newly created VPN to access the 4-Degree Consulting server as well as the dialog box labeled Connect that is activated if you click on the VPN icon shown in the background of the figure.

Note that the cursor on the Connect dialog box is over the button labeled Properties, which provides us with the ability to view and, if necessary, change default values associated with the previously created connection. Thus, let us click on the Properties button and examine our options.

Figure 6.15 illustrates the Properties dialog box for the previously created VPN we named 4-Degree Consulting. Note that this dialog box includes five tabs, with the General tab shown positioned in the foreground.

General tab

The General tab provides you with the ability to change the IP address or host name of the server with which you wish to form the VPN. In addition, this tab allows you to specify or change a telephone number to be dialed to connect to an Internet Service Provider prior to establishing the VPN connection. Thus,

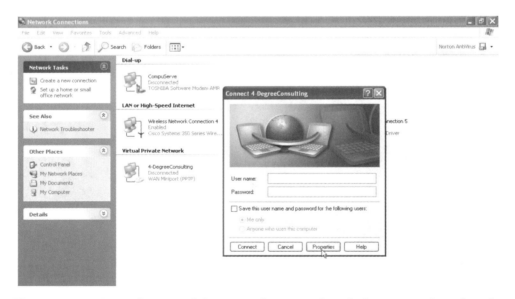

Figure 6.14   Once the use of the wizard is completed, the VPN is listed in the Network Connections screen and activated by double clicking on the entry.

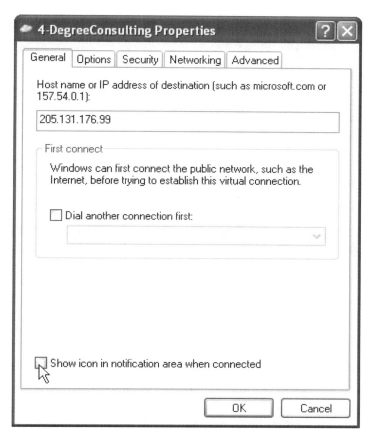

Figure 6.15 The General tab provides you with the ability to alter the server destination and specify a dial connection telephone number.

it is possible to create a configuration for use at home to connect to your organization's server via a DSL or cable modem and modify that configuration when 'on the road' for dial-up use. To do so, you would enter a telephone number for a dial connection through the use of the General tab in the VPN's Properties dialog box.

Prior to moving on to the next tab, turn your attention to the lower left portion of Figure 6.15. Note the cursor is shown placed on a check box which, when selected, will generate an icon in the notification area of the display when you have a VPN connection. Thus, selecting this option provides a visual indication of your VPN connection that can be important when you are manipulating multiple windows on your display.

Options tab

The second tab in the VPN Properties dialog box is the Options tab. Figure 6.16 illustrates the Options tab placed in the foreground of the VPN Properties dialog box, showing its default settings.

Note the Options tab is divided into two sections, both of which are focused on telephone dialing. The top section consists of three dialing options, of which two, display progress and prompt for name and password, are selected by default. The lower section of the Options tab has three re-dialing options and a check box setting to activate re-dialing. By default re-dialing is disabled. By clicking on the check box you can activate re-dialing if the line providing a connection to the Internet is dropped.

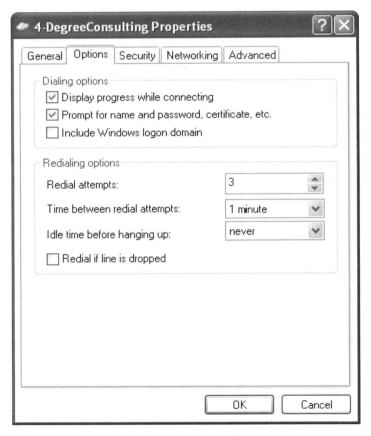

Figure 6.16 The Options tab provides the ability to control both dialing and re-dialing.

As indicated in the re-dialing options section, by default your configuration is set for three re-dial attempts using a one-minute period between re-dials without hanging up the line. Using the pull-down menus associated with each option, you can change the default settings to another pre-defined value.

Security tab

Because VPNs are primarily used to provide security, it should come as no surprise that the Security tab has a number of settings that correspond to various security features associated with different VPN protocols.

Figure 6.17 illustrates the default settings associated with the VPN Properties Security tab. By default authentication occurs by requiring a secure password and data encryption is required to protect data from being examined and understood by unauthorized third parties. If your organization uses a smart

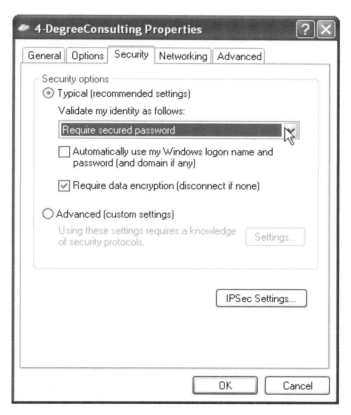

Figure 6.17 The Security tab enables the selection of different types of authentication and encryption.

card for authentication you would use the drop down menu pointed to by the cursor in Figure 6.17 to select a different validation method. If you want to automatically use your Windows logon name and password you would select the checkbox to the left of that entry.

Two additional options on the Security tab that deserve mention are the Advanced button, which when clicked on allows you to select the Settings button, and the IpSec Settings button.

Advanced Security Settings

The selection of the Advanced radio button in Figure 6.17 enables you to select the Settings button associated with that option. Doing so results in the display of the Advanced Security Settings dialog box, which is shown in Figure 6.18.

Figure 6.18 The Advanced Security Settings dialog box provides settings to control encryption, the use of a smart card or digital certificate and the selection of basic authentication protocols.

Under Windows XP there are four data encryption options available for client selection. Those options are listed below, with the top option representing the default setting:

- Require encryption (disconnect if server declines).
- No encryption allowed (server will disconnect if it requires encryption.
- Optional encryption (connect even if no encryption).
- Maximum strength encryption (disconnect if server declines).

Selecting the second option in effect defeats the primary purpose of creating a VPN, but is probably offered for computer platforms that may be hard-pressed to provide full functionality while encrypting and decrypting data. Similarly, selecting the optional encryption entry is not advisable if you are attempting to protect sensitive data transmitted to and from the workplace server.

If you select the radio button associated with EAP under the logon security section of the display, you can select either the use of 'MD-Challenge or Smart Card' or 'other Certificate' for logon security. When you select EAP you cannot use any of the protocols listed in the lower portion of Figure 6.18. Concerning those protocols, by default Microsoft's MS-CHAP and MS-CHAPv2 are selected for authentication. As indicated previously in this book, the use of PAP results in the transmission of passwords in the clear. Thus, Microsoft notes this by including the prefix 'unencrypted password' prior to PAP. Because it is relatively easy for an unauthorized third party to read a PAP password and use it to gain access to a computer, it is probably best to stick with the Microsoft default settings unless you intend to use a stronger method supported by EAP.

Returning our attention to the log on security section of the Advanced Security Settings dialog box shown in Figure 6.18, when you select the radio button associated with EAP you also obtain the ability to select the Properties button associated with that entry. When you select the Properties button you obtain the ability to specify the use of a smart card or certificate as well as to select one or more trusted root certification authorities and view different certificates. Figure 6.19 illustrates the dialog box entitled 'Smart Card or other Certificate Properties'. Because IPSec uses a certificate for computer authentication you would select the certificate radio button if you intend to use that security method. Because versions of Windows prior to Windows 2000 do not support IPSec natively, unless your organization is operating third party software to obtain an IPSec capability you would not want to use a certificate for log on when establishing a connection to the server.

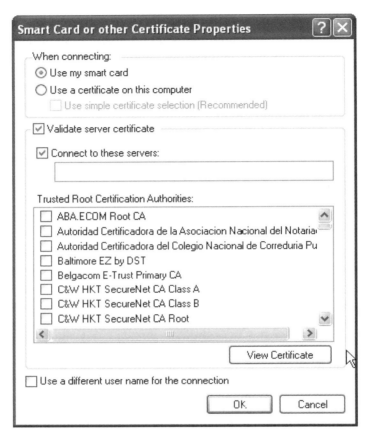

Figure 6.19 The Smart Card or other Certificate Properties dialog box pro-vides the ability to restrict connections to specific servers and view the contents of certificates.

IPSec Settings

Returning our attention to the IPSec Settings button shown in Figure 6.17, clicking on that button provides you with the ability to define a pre-shared key to be used for authentication. Figure 6.20 illustrates the IPSec Settings box, showing the entry of a portion of this author's famous 'abadabado4you' entry.

If you compare the IPSec settings on the client side to the settings on the server side previously examined earlier in this book, it is obvious that the server provides considerable more tailoring. As an educated guess for the reason for the discrepancy between client and server side IPSec configuration capability, this author believes there are actually two. First, Microsoft more

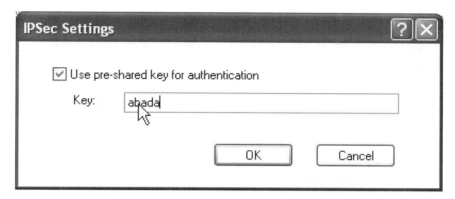

Figure 6.20 The IPSec Settings dialog box permits the use of a pre-shared key for authentication.

than likely attempted to hide the complexity of the IPSec configuration from the wide range of client side users. Thus, keeping IPSec as simple as possible for clients to use by only having to enter a pre-shared key for authentication makes the configuration process easy. Second, Microsoft's IPSec implementation automates a configuration exchange based upon the server side settings once the client's pre-shared key setting is validated. Between the design process and the manner by which IPSec client – server communications occur, client side configuration is kept to a minimum.

Networking tab
Continuing our journey through the various tabs in the VPN Properties dialog box, Figure 6.21 illustrates the Networking tab. While most of the default settings are sufficient for connecting to a modern Windows server, a few words are in order concerning the pull-down menu associated with the Type of VPN. The default setting of Automatic obviously works well with a Microsoft Windows server. The two additional entries available from the pull-down menu include PPTP VPN and L2TP IPSec VPN. If you are connecting to a third party server you may need to change the 'Automatic' setting to a specific setting. However, if you are connecting to a Windows 2000 server running Microsoft Routing and Remote Access (RRAS) Server, the Automatic setting is acceptable, as the client will adjust its VPN to the configuration of the server.

The lower portion of the Networking tab provides a summary of the protocols that will be used for the connection. By default, TCP/IP, a Quality of Service (QOS) packet scheduler and the client for Microsoft Networks are used and

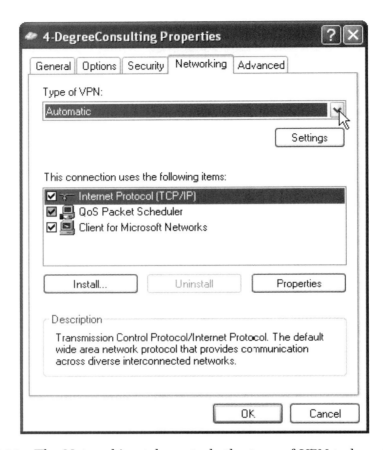

Figure 6.21   The Networking tab controls the type of VPN to be established and the protocols used by the connection.

shown in the lower portion of Figure 6.21. You can examine or modify the properties of any protocol used by the connection by clicking on the button labeled Properties. If for some reason you need to add an additional protocol for your VPN connection you can use the Install button to add one or more protocols and then tailor the addition or additions through the use of the Properties button.

Advanced tab
The fifth and last tab in the VPN Properties dialog box is labeled Advanced. The purpose of this tab is twofold. First, it provides a mechanism to configure Microsoft's Internet Connection Firewall. Although the Internet Connection

Firewall is enabled by default, the only option pre-configured is a log file for logging either dropped packets or successful connections, with an adjustable size limit configured for the file. Through the use of a Settings button on the Advanced tab, you obtain the ability to establish filters that enable a specific host or IP address using a pre-defined TCP or UDP port number to access the client. In addition, through the Internet Connection Firewall you can also enable four incoming and four outgoing ICMP messages as well as redirect ICMP messages.

The second function of the Advanced tab is to control Internet connection sharing. By default, this option is disabled. However, through its use you can allow other network users to connect through the VPN you are configuring. Once you complete your configuration of the various properties associated with the VPN connection, you can either use the connection or return to the Connect dialog box previously shown in Figure 6.14. From that dialog box you can enter a user name and password that can be saved for future use for yourself or for anyone who uses the computer. While the latter may appear risky, under Windows XP you can protect each user's access to the computer with a password. Because it is only a matter of time before Windows XP password cracking programs are readily available on the Internet, it is a good idea to use a long alphanumeric string for each password. Doing so eliminates the possibility of a dictionary attack as well as substantially increases the time required for a brute force attack. Concerning the latter, restricting passwords to just uppercase characters and digits results in each additional password position increasing the number of combinations that must be tried under a brute force method by the power of 36.

Now that we have an appreciation of the client-side VPN configuration process under Windows XP, let us turn our attention to the server side. In doing so we will examine the configuration of a Windows 2000 server.

6.3 Working with Windows 2000 server

With the release of Windows 2000, Microsoft provided users with an alternative to the use of PPTP and some of the potential security problems associated with its use. That alternative was in the form of support for the Layer 2 Tunneling Protocol (L2TP) which uses IPSec for data encryption. In this section we will turn our attention to the configuration of a Windows 2000 server for VPN support. Because you need to install Microsoft's Routing and Remote Access Service (RRAS) prior to being able to tailor the configuration of a VPN, we will first examine the installation of RRAS in this section. Once this is done, we will then turn our attention to the VPN configuration process, with special attention focused upon PPTP and L2TP combined with IPSec.

6.3.1 Installing RRAS

Under Windows 2000 you must be logged on as the administrator or as a member of the administrator's group to be able to install and configure the remote access server. Because RRAS is designed to provide a termination service for remote users accessing the server, you first need to install applicable hardware, such as a network adapter and/or one or more ISDN adapters or modems. Once this is accomplished you can then install RRAS.

6.3.2 Enabling RRAS

In reality, when you install a Windows 2000 server, its remote access facility is automatically installed. However, Microsoft's Routing and Remote Access Service is installed in a disabled state. Thus, you need to enable RRAS. You can enable RRAS by clicking on Start/Programs/Administrative Tools, then selecting Routing and Remote Access. This action will activate the Microsoft Management Console (MMC) window, with your server listed in the left portion of the window similar to Figure 6.22, with the right portion

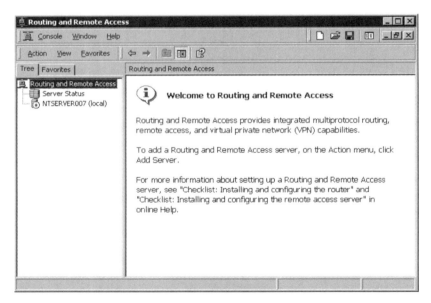

Figure 6.22 Routing and Remote Access Service is activated through the Microsoft Management Console.

of the MMC displaying information about RRAS. In this example the author's server was labeled in honor of a famous British character you will obviously recognize if you are a fan of Ian Fleming.

By default, the local computer is listed as a server and we will shortly configure that server. In comparison, if you wanted to enable RRAS on another server you could either right click on the Server Status entry in the left portion of Figure 6.22 or use the action menu to select Add Server, either action displaying a dialog box allowing you to specify a particular computer.

To set up RRAS, on the Action menu in the MMC you would click on a drop-down entry labeled 'Configure and Enable Routing and Remote Access.' Doing so begins the set up process by activating another Microsoft wizard. As you might expect, this wizard is titled 'Routing and Remote Access Server Setup Wizard.'

6.3.2.1 Using the Wizard

Figure 6.23 illustrates the initial RRAS wizard screen display. There are five radio buttons on this initial screen, each associated with a configuration option.

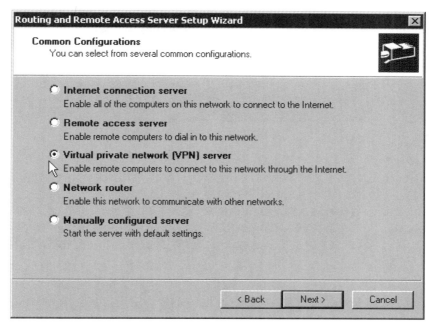

Figure 6.23 The Routing and Remote Access Server Setup wizard supports five common configurations.

The initial step associated with the use of the configuration wizard is to determine the role your server will play with respect to servicing communications or performing routing. For example, if you wanted to accept connections from dial-in clients, you would select the radio button to the left of the Remote Access Server label. Because we want to configure the server to support remote computers communicating through the Internet, we selected the third radio button. As indicated in Figure 6.23, that button is used to select a configuration for a VPN server.

Remote Client Protocols

After you select the VPN Server radio button and click on the button labeled Next the wizard will display a dialog box labeled 'Remote Client Protocols.' An example of this dialog box is shown in Figure 6.24.

The purpose of this screen is to allow you to verify that the protocols installed on your server match the requirements for providing remote access to your VPN clients. The only correct answer to this dialog box is yes, since clicking the radio button associated with the No entry will not allow you to add other protocols. Instead, clicking the radio button associated with No and

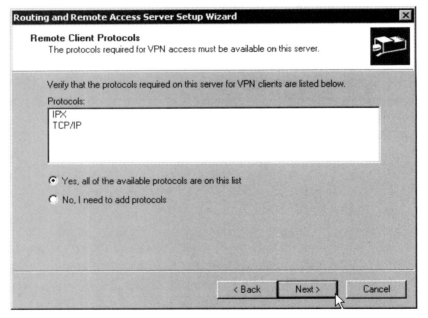

Figure 6.24 The Remote Client Protocols screen indicates the protocols available on the server.

clicking on the button labeled Next will display a 'cannot continue' message. The message will inform you that you need to add the required protocols from the Network and Dial-up Connections folder and then re-run the wizard. Because our remote clients will access the server via the Internet, we need the TCP/IP protocol, which is shown installed. Thus, we would check the top radio button and then click on the button labeled 'Next' to access the next wizard screen.

Internet Connections

The following wizard screen is labeled 'Internet Connection'. The purpose of this screen is to specify the Internet connection that the server uses. An example of this screen is shown in Figure 6.25, with the highlight bar shown placed to indicate that the local area connection will be selected as the mechanism which provides access to the Internet.

In reality, one of the minor puzzles of RRAS that takes some time to understand is that the Internet connection referred to by Microsoft is to specify a

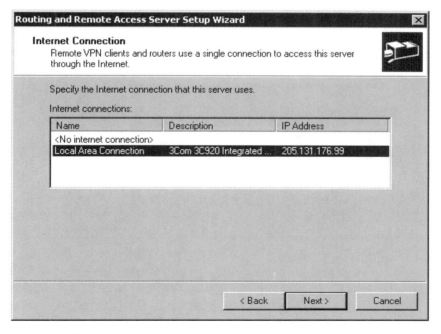

Figure 6.25 When using the Internet connection dialog box you need to select the 'No internal connection' entry if your server is connected to the Internet via a LAN.

Dataphone Digital Service (DDS), T1 or fractional T1 connection terminating in a WAN adapter. Thus, if you attempt to specify the local area network connection, RRAS will generate the cryptic warning message 'You have chosen the last available connection as the Internet connection.' A VPN server requires that one connection be used as the private network connection. What this means is that the server used by this author was connected to a LAN which in turn was connected to the Internet. Thus, there was no direct Internet connection, requiring the selection of the top entry (no Internet connection) in the dialog box in order to continue using the wizard. Perhaps a better name for the top entry in the dialog box shown in Figure 6.25 would be 'no direct Internet connection.'

IP address assignments

The next step associated with using the wizard occurs if you have previously installed TCP/IP on your server. That step concerns IP address assignments. By default, the RRAS setup wizard selects a radio button associated with the automatic configuration of IP addresses via a DHCP server. As an alternative, you can select IP address assignments to occur by specifying a range of addresses. Figure 6.26 illustrates the IP address assignment dialog box.

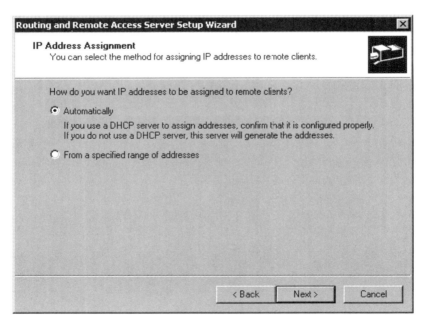

Figure 6.26   You need to assign IP addresses to remote dial-in users to provide them with a mechanism to participate on the server's network.

IP address assignments result in the server issuing a temporary address to a remote client. This action is designed so that each device on the VPN network has its own IP address and enables direct dial-in users to be supported. Because we are in the process of configuring the server to accept remote clients with IP addresses, this wizard screen is really not applicable for our use and we can simply accept the default and click on the button labeled Next. As an alternative, if you also anticipate supporting remote dial-users and do not wish to use DHCP, you could specify the second radio button in Figure 6.26. This action would result in the display of a dialog box labeled 'New Address Range.' You would then enter the start and end IP addresses into this box to specify a new address range. In fact, Windows 2000 server permits you to enter additional address ranges, with the server assigning all of the addresses in the first range prior to assigning addresses from the next range.

RADIUS

Once you pass the IP address assignment display the wizard will provide you with a mechanism to handle authentication and logging via a Remote Authentication Dial-In User Service (RADIUS) instead of using the RRAS authentication and logging facility. If your organization is currently using a RADIUS server, you would probably prefer to use that server to centrally manage authentication. In this situation you would select the second radio button shown in Figure 6.27 and click on the Next button.

If you want to use RRAS for authentication and logging, you would select the top radio button, which is the default setting. Because the option for selecting the use of a RADIUS server is the final step in the use of the wizard, clicking on the Next button results in the display of a 'completing' screen which informs you that you have successfully completed the configuration of a VPN server. You can then close the wizard by clicking on the new screen's Finish button.

6.3.3 Configuring RRAS

Now that we have enabled RRAS, let us turn our attention to its configuration to support PPTP and L2TP with IPSec. In doing so we need to note that the use of L2TP with IPSec requires the use of the Encapsulating Security Payload (ESP), Internet Key Exchange (IKE) and L2TP. We also need to note the ports used by these protocols, since the placement of the server behind a router operating an access list or a firewall will require a reconfiguration of those devices to support data transfer on certain ports.

Because PPTP is transported via Generic Routing Encapsulation (GRE) packets, you need to ensure that IP protocol 47 is open at your router or

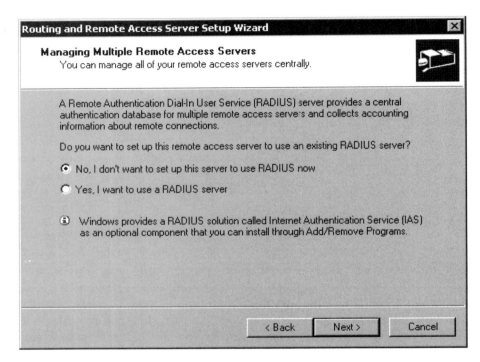

Figure 6.27 You can either use RRAS authentication or specify the use of a RADIUS server.

firewall in front of your server. In addition to GRE Protocol 47, TCP port 1723 must be enabled at all routers or firewalls between the PPTP client and server. To support L2TP and IPSec you need to allow the flow of data for ESP, IKE and L2TP. Those ports include UDP port 1701 for L2TP and UDP port 500 for IKE. For ESP, which is transported by IP, if our devices are filtering on the IP protocol number we need to allow IP protocol 50. Table 6.3 summarizes the ports and protocol IDs required for each type of VPN.

TABLE 6.3 Ports and protocol IDs

VPN	Ports and protocol IDs
PPTP	TCP port 1723
	IP protocol 47
L2TP with IPSec	UDP port 500
	UDP port 1701
	IP protocol 50

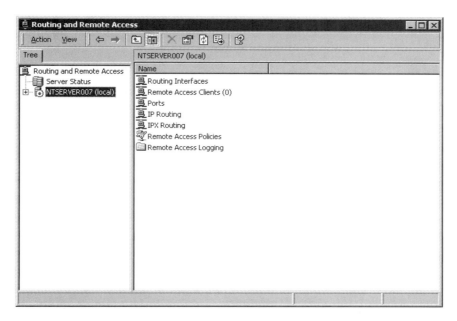

Figure 6.28   Selecting the local server for RRAS configuration.

To configure Routing and Remote access Server (RRAS) we would start the MMC for Routing and Remote Access and select the server we wish to support. The resulting display would be similar to that shown in Figure 6.28 for the server used by this author. That server, labeled NTSERVER007, is shown highlighted in the left portion of Figure 6.28.

6.3.3.1 Configuring the Interface

To configure the interface that will accept the VPN connection, you would expand the selected server object, then expand the resulting display of the IP routing object, selecting the entry labeled 'General' under IP Routing. This action is illustrated in the left window of Figure 6.29. Note that the right window in Figure 6.29 lists the available interfaces on the server. Because we want to configure the LAN interface to accept VPN connections we would double-click on that interface to examine its properties as well as to initiate applicable configuration changes to support PPTP and L2TP with IPSec connections.

Figure 6.30 illustrates the connection properties dialog box for the local area connection we previously selected in Figure 6.29. In the General tab, which is

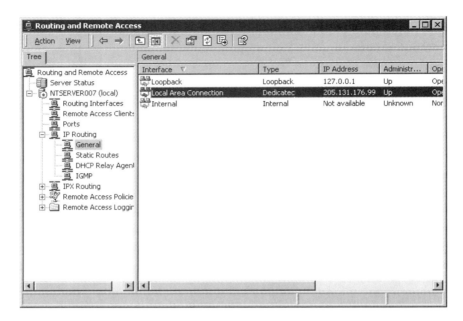

Figure 6.29 To configure an interface for VPN support, explore IP Routing in the RRAS dialog box and select the General entry. Then select the interface in the right window.

shown in the foreground of Figure 6.30, you will note two buttons in the lower left portion of the box. Those buttons, which are labeled 'Input Filters...' and 'Output Filters...' provide us with the ability to control which packets can be received or transmitted on the previously selected interface.

The creation of filters is optional, however, their use can enhance security as they control the flow of packets in the inbound and outbound direction with respect to the server. Because an improper or ill-conceived filter can block the flow of intended traffic, you should carefully review any filters you create. To illustrate the use of the filter let us click on the Output Filters button.

Filter control

Clicking on the button labeled Output Filters results in the display of a dialog box, which provides you with the ability to create applicable filters to support packet flow control for a VPN. As previously noted, you can control access to the VPN server based upon source and destination IP addresses contained within a packet as well as the source and destination port numbers in the packet.

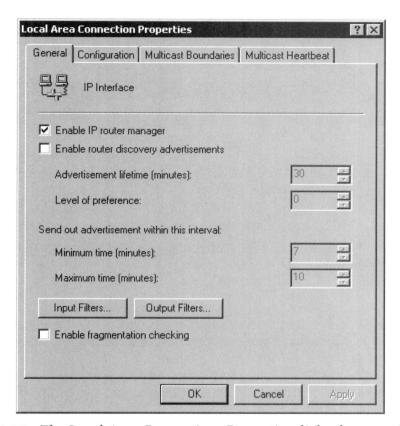

Figure 6.30 The Local Area Connections Properties dialog box provides the ability to control packets received for forwarding or processing by IP address and port number.

The Output filters dialog box is shown in the left portion of Figure 6.31. Note that the window labeled 'Filters' is presently blank as we have yet to create any. Clicking on the button labeled 'Add' results in the Add IP Filter dialog box being displayed which is shown in the right portion of Figure 6.31.

Assuming the IP address of our server is 205.131.176.99, we can restrict packets flowing from the server to those generated by the server by using a mask of 255.255.255.255. Because this author's home IP address is 68.107.198.255, a mask of 255.255.255.255 is shown for the destination network. If we wanted to support the flow of packets to multiple computers on the destination network, we would adjust the subnet mask for that network accordingly. For example,

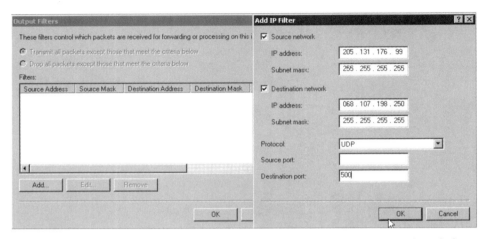

Figure 6.31 The Output Filters dialog box provides you with the ability to control packets transmitted by the server based upon their IP addresses and port numbers.

a subnet mask of 255.255.255.0 would allow packets to flow to all devices on the 68.107.198.0 network. Concerning the protocol, to support IKE you would select UDP and then enter 500 for the destination port.

Once you add the filter shown in the right side of Figure 6.31, you need to add a filter for L2TP traffic. That filter for our example would have the same source and destination addresses, however, the destination port value would be changed to 1701 for L2TP. We do not have to specify output or input filters for ESP traffic (IP Protocol 50) because IPSec removes the ESP header prior to the enforcement of RRAS filters.

Figure 6.32 illustrates the Output Filters dialog box after we entered two filters. Because we want to limit packet flow to those meeting our previously defined criteria, we selected the second radio button at the top of the window. Note that selecting the top radio button would not allow us to communicate with the destination address.

If we did not enter a destination address or subnet mask for a filter, the display would indicate 'any' for each entry. This action would provide more flexibility, since it would enable the server to communicate via L2TP with IPSec to any client regardless of its IP address. On the downside, it removes an addressing restriction that makes communications more secure.

Once you configure the Output filters you need to consider specifying a series of input filters, if you wish to control packets the server will accept for forwarding or processing. In doing so, remember that the source address

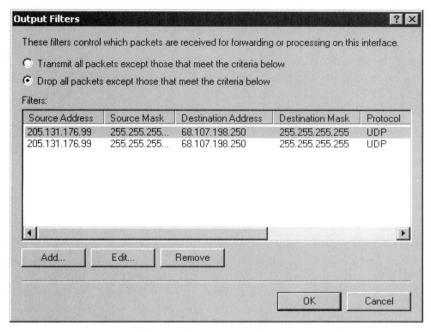

Figure 6.32   The Output Filters dialog box after adding filters to support IKE and L2TP packets flowing to a specific IP address.

now represents the remote client or clients while the destination address represents the server or stations on the server's network depending upon the composition of the destination mask.

When setting up input filters you need to be very careful to specify criteria for the VPN methods you wish to support. For example, if you wish to support both PPTP and L2TP with IPSec access, specifying input filters that restrict access to L2TP with IPSec will result in PPTP clients receiving the following error message: 'Error 701: The encryption attempt failed because no valid certificate was found.' This error message was generated because this author configured a client to access the server with client side configured for 'automatic' when defining the type of VPN server being accessed. During negotiation the client could only access the server on ports enabled for L2TP with IPSec, which requires the use of certificates for authentication. Because the client did not have a certificate, the previously noted error message was displayed. However, if ports for both PPTP and L2TP with IPSec were allowed in applicable input filters, the client would have been able to connect to the server via PPTP.

6.3.3.2 Authentication Requirements

If your client wants to connect to the server using PPTP, authentication can occur via PAP, CHAP or MS-CHAP. In addition, you can select the use of EAP and use either a smart card, an MD-5 challenge or a certificate. If your server supports L2TP with IPSec, authentication can occur either via certificates or through the use of Microsoft's integrated Kerberos.

To ensure your server has a computer certificate available for IPSec utilization you would start the Microsoft Management Console by selecting Start, Run and typing MMC on the command line. The MMC is shown in the background of Figure 6.33.

From the MMC you would select the Add/Remove-Snap-in from the Console menu. The resulting dialog box is partially shown superimposed on the MMC in the previously referenced figure. Clicking on the Add button would result in the display of the Add Standalone Snap-In dialog box, which is shown as the third window from the left. You would select the Certificates entry from that box and click on the Add button, resulting in the display of the

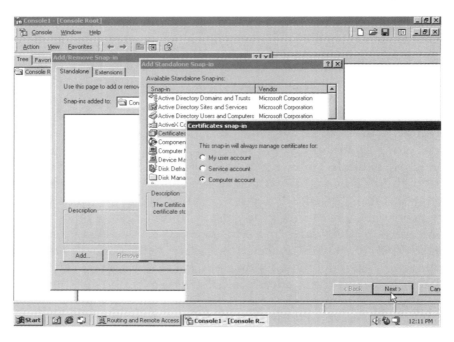

Figure 6.33   You can view available certificates through the MMC by selecting the Certificates entry from the Add/Remove Snap-in.

Certificates snap-in box shown in the foreground of Figure 6.33. Next, you would select the 'Computer account' radio button and click on the button labeled Next. This action will result in the display of a dialog box, which enables you to specify the computer for the snap-in to manage. After you specify the computer, you can return to the MMC and view the certificates available for use. To support IPSec a certificate must:

- be stored in the computer account;
- contain an RSA public key that has a corresponding private key that can be used for RSA signatures.
- be used within the certificate validity period.
- have a trusted root certification authority.

Once you verify the availability of an applicable certificate and, if necessary, obtain one, you need to verify applicable ports are enabled to support your server VPN connections. Thus, let us turn our attention to this topic.

6.3.3.3 Port Configuration

The ability to support PPTP and L2TP connections requires your server to have applicable ports enabled to support each type of VPN. You can verify the availability of PPTP and L2TP ports via RRAS, expanding the server entry and selecting the Ports object under the server.

The left portion of Figure 6.34 illustrates the selection of the Ports object for NTSERVER007. By right clicking on the Ports object and selecting Properties from the pop-up menu, the dialog box shown in the right portion of Figure 6.34 is displayed. Note that by default RRAS provides 128 PPTP and 128 L2TP miniports. You can double-click on either entry to configure PPTP and L2TP ports for RRAS access, with the resulting dialog box only varying concerning the label of the box.

Figure 6.35 illustrates the dialog box for configuring a WAN miniport for PPTP. As previously noted, the dialog box displayed if you clicked on the L2TP entry in the right window of Figure 6.34 would replace PPTP by L2TP.

For both PPTP and L2TP the default setting has ports configured for Remote access connections and Demand-dial routing and most users should retain those settings. Since our example will result in a client accessing the server via a cable modem connection to the Internet, we do not need to specify a phone number. Because this author will later modify the previously created filters to enable additional remote access but wants to limit the maximum number of remote clients, he will modify the maximum ports setting. Instead of the

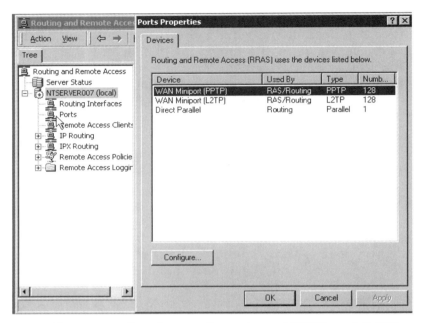

Figure 6.34 You can tailor the configuration of PPTP and L2TP ports by double clicking on the Ports object for the selected server.

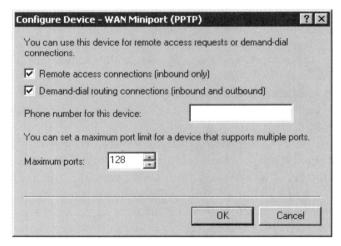

Figure 6.35 Through the Configure Device box for applicable VPNs you can control data flow, assign a telephone number and control the maximum number of ports supported by PPTP and L2TP.

default value of 128 for PPTP and L2TP this author lowered the maximum port limit to 5 of each type of VPN. Later in this section when we return to viewing server ports we will see the results of this port reset operation.

6.3.3.4 Viewing Server Properties

Prior to being able to have remote clients access our server we need to consider three additional areas. Those areas are server properties, RRAS and IP Security policies, and user accounts. We can view server properties by right clicking on the server object in the RRAS console and selecting properties from the resulting pop-up menu.

Routing and Remote Access

Figure 6.36 shows the properties dialog box for the server we are configuring, with its General tab positioned in the foreground. By default, the server's General tab is configured to support LAN and demand-dial routing and a remote access server capability and you should consider retaining these settings.

If you disable the check box to the left of the Router entry you will disable the server's ability to support routing, preventing clients from accessing other computers on the server's network via the server. Removing the check to the left of the 'Remote access server' entry prevents the server from supporting remote clients. If you enable routing but disable the remote access server capability, the server will be limited to supporting remote clients accessing other computers on the server's network.

Verifying Authentication

To verify server authentication settings you need to click on the Security tab in the server properties dialog box and click on the button labeled 'Authentication.' Doing so will result in the display of a dialog box labeled 'Authentication Methods,' which is shown in Figure 6.37.

In examining the entries shown in Figure 6.37, it is important to note that as the text on the display mentions, the server authenticates remote clients by using the selected methods in the order presented, from top (most secure) to bottom (least secure). To support L2TP you need to ensure the box next to EAP is checked and then click on the button labeled 'EAP Methods' to select the method your server supports. For PPTP most organizations will more than likely use MS-CHAPv2 or MS-CHAP, which are set along with EAP by default. If your organization uses a different authentication method, you should check the applicable box to provide support for that method.

At the bottom of Figure 6.37 you will note a checkbox associated with unauthenticated access. Because a VPN is used for security you might be puzzled

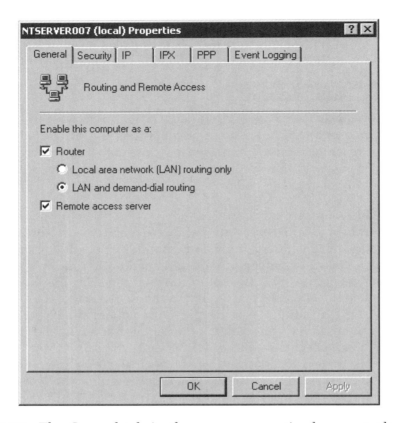

Figure 6.36 The General tab in the server properties box controls routing and remote access.

as to why an organization might wish to allow access from remote computers without authentication. One example could be an organization that is concerned about privacy of data and not security. By constructing appropriate inbound filters you could restrict access to pre-defined IP addresses, allowing remote clients on one LAN to access a server via the Internet. Because the server might be limited to providing parts or sales delivery information, there is no need to authenticate remote users.

6.3.3.5 RRAS and IP Security Policies

Continuing our effort to configure the server for remote VPN access, we would return to the RRAS window in the MMC and review the creation of a Remote Access Policy to reflect our organizational requirements. For

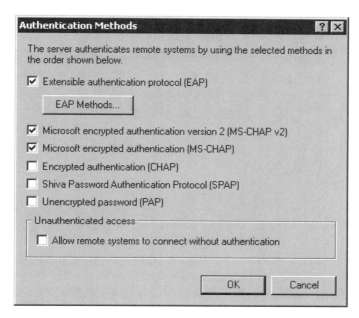

Figure 6.37 A Windows server authenticates remote systems by using selected methods in a pre-defined order.

some organizations the use of filters and the creation of user accounts with applicable permissions might be sufficient. In comparison, other organizations may prefer to create a RRAS Remote Access Policy consisting of one or more attributes.

To create a RRAS Remote Access Policy you would first right click on the Remote Access Policies object, which is shown, highlighted in the window positioned in the background of Figure 6.38. Next, you would select New Remote Access Policy and name the policy. Clicking Next will result in the display of the middle window shown in Figure 6.38. This window is currently blank, as we have not yet specified any attributes for the policy. To select one or more policy attributes click on the Add button. This action results in the display of the Select Attribute box shown in the foreground of Figure 6.38. From this box you would add applicable attributes to form your policy.

If you turn your attention to the Select Attribute dialog box shown in the foreground of Figure 6.38 you should note how you can create a policy that can be very restrictive concerning the access capability of a remote user. For example, you can restrict access to a particular dialed or dialing phone

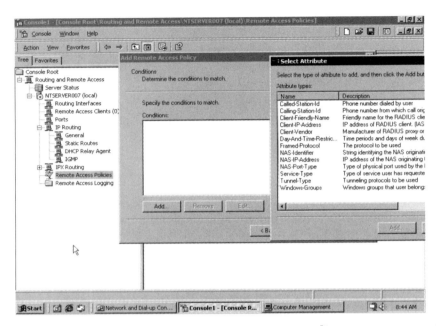

Figure 6.38   You can tailor a RRAS Remote Access Policy to your operational requirements.

number, for a pre-defined time period or for membership in a Windows group to which the user belongs.

After you create a RRAS Remote Access Policy you need to consider the creation of an IP Security policy on the RRAS server. If you intend to support L2TP with IPSec you need to require IPSec between the RRAS client and the server, which can occur through the creation of an IP Security Policy.

To create an IPSec Policy on the RRAS server you would first start an MMC console. You would then add the snap-in for IP Security Policy Management for the local computer and double click on the server policy entry. This action would result in the display of a Secure Server Properties dialog box, similar to the window shown in Figure 6.39.

Note that this window has two tabs, labeled Rules and General, with the Rules tab positioned in the foreground. That tab has three pre-defined IP Security Rules.

To establish an IP Security policy you would double click on the top entry in the IP Security Rules Window which is labeled 'All IP Traffic.' This action would result in the display of a dialog box labeled IP Filter list, which is illustrated in Figure 6.40.

Figure 6.39 To create an IP Security policy you would double click on the All IP Traffic entry in the Secure Server Properties dialog box.

When examining the IP Filter list dialog box shown in Figure 6.40, note that it consists of three windows and a series of buttons. The top window denotes the name of the filter and is shown as 'All IP Traffic' since we selected that entry from the Secure Server Properties box previously shown in Figure 6.39. The middle window provides a description of the filter, while the bottom window lists previously created filters.

If you look at the bottom window in Figure 6.40 you will see that by default one filter exists. That filter results in the matching of all IP packets from the server to all other computers, with the exception of broadcast, multicast, Kerberos, RSVP and IKE packets. Thus, the entries in the single line in the lower window are all set to 'ANY.'

From the IP Filter list you would select the default filter in the lower window shown in Figure 6.40. Once this is accomplished, you would double click on

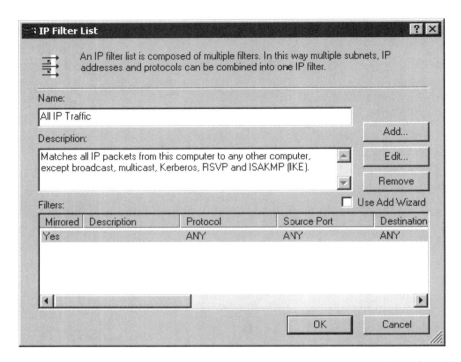

Figure 6.40 Through the IP Filter list you can view, create and modify filters that will affect the flow of IP traffic.

that entry, resulting in the display of a dialog box labeled Filter Properties. That box has three tabs labeled Addressing, Protocol and Description. You would select the Protocol tab as shown in Figure 6.41.

From the Protocol tab you would select UDP from the top drop-down box. Because L2TP occurs using UDP port 1701, you would select the radio button associated with 'To this port' and enter 1701 as the 'to this port' value. Once this action is accomplished you can click on the Apply and OK buttons and click on a series of Close buttons to return to the IP Security Policy console.

6.3.4 Creating a test account

Although it is assumed that remote clients that will access the server via a VPN connection have an account on the server, it is important to note that not all accounts are created equal. Another important item to consider is the difference between RRAS remote access policy and a user account. As previously noted, a RRAS remote access policy can tailor VPN access

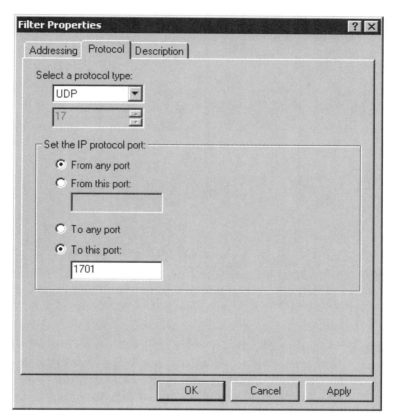

Figure 6.41 To restrict VPN access to L2TP you would filter on UDP port 1701.

to specific originating and/or destination telephone numbers, control access by day and even time of day, type of service requested and other attributes which were previously shown in the window positioned in the foreground of Figure 6.38. In comparison, establishing a user account provides you with the ability to assign and control a password, associate a user with one or more groups which in turn govern the ability of the user to perform such actions as reading, writing and modifying files. In addition, you can assign a logon script to a user as well as enable or disable VPN access for a user.

Figure 6.42 shows the use of the MMC to create an account appropriately named test account.

Note that the test account properties box located in the right portion of Figure 6.42 has four tabs, with the tab labeled 'General' shown in the

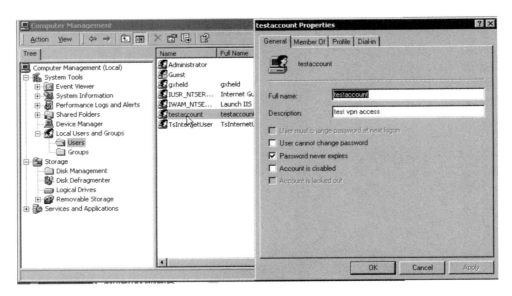

Figure 6.42 Creating an account for testing VPN access.

foreground. The 'Member of' tab provides you with the ability to assign the user to one or more pre-defined accounts, such as Administrators, Backup Operators, and so on. By default, a newly created account is a member of the Users group. The Profile tab provides the ability to assign a log on script to the user as well as create a pre-defined location to which they are placed when they remotely access the server, such as a local path or connection to a network drive. The fourth tab, which is labeled Dial-in, is primarily used for dial-in connections and controls such items as caller verification and the assignment of a static IP address. However, this tab also allows you to control both dial-in and VPN remote access. In doing so, the Dial-in tab provides you with the ability to allow, deny or control access through a Remote Access Policy. Thus, if you have previously created a Remote Access Policy, you need to select the radio button on the Dial-in tab of each account properties box associated with controlling access through Remote Access Policy.

6.3.5 Testing the connection

Now that we have established a test account, let us test our connection. Figure 6.43 illustrates a Windows client Connect Virtual Private Connection dialog box for which we assigned the User name 'testaccount' and the applicable password.

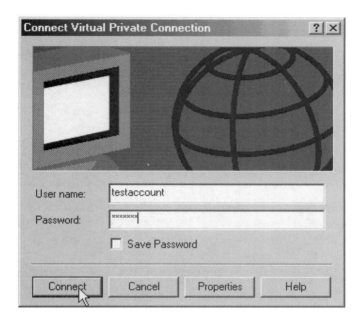

Figure 6.43 Preparing for a VPN Connection using 'testaccount' for the user name.

Previously, we used the Properties button to define the IP address of the server with which we wish to connect, set dialing options to display progress while connecting, used a Security tab in the dialog box to define the use of data encryption, and authentication protocols and the Networking tab to define the type of VPN server with which we wish to communicate. Borrowing an old adage from a former professor, since the proof of the pudding is in the eating, let us click on the Connect button shown in the lower left portion of Figure 6.43 and establish a connection to the server.

Because access occurred from this author's client station via a cable modem connection to the Internet and the server was only a few router hops away, the connect progress display rapidly flashed off the screen. That display was replaced by the Connection Complete box, which is shown in Figure 6.44.

That box informs you that you can check on the status of your VPN connection by right clicking on its icon in the Network Connections folder or, if you configured your client to display an icon on the task bar, you can simply click on that icon. As a refresher, on all versions of Windows this icon consists of two miniature computers similar to the icon in the upper left portion of Figure 6.44. Each computer will have a green flash, which corresponds to

Figure 6.44   Once a client establishes a VPN connection, the Connection Complete box will be displayed.

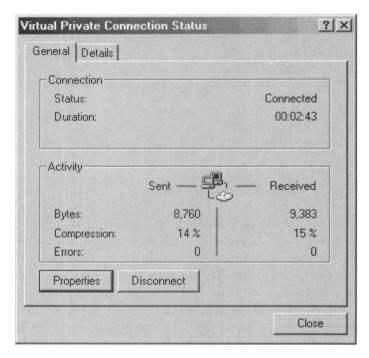

Figure 6.45   The Virtual Private Connection Status box can be used to observe parameters associated with your VPN connection as well as to view and change previously defined properties.

the transmission of a packet. Clicking on the icon will result in the display of a Virtual box labeled Private Connection Status. Figure 6.45 illustrates an example of the Virtual Private Connection Status dialog box at a period in time when the remote client had been connected for a period of two minutes and forty-three seconds with the server.

In examining the Virtual Private Connection Status box shown in Figure 6.45 you will note that the box has two tabs, with the General tab shown positioned in the foreground. From this tab you can disconnect a previously established connection by clicking on the button appropriately labeled Disconnect. In addition, you can view and, if necessary, modify previously established properties by clicking on the button with that label. This action would display the dialog box labeled Virtual Private Connection that contains the tabs General, Options, Security, Networking, and Sharing. If you select the tab labeled 'Details,' you will be able to view details concerning your VPN connection. Those details include the server type supporting your connection, the methods used for authentication, encryption and compression, whether or not PPP multi-link framing is enabled and client and server IP addresses.

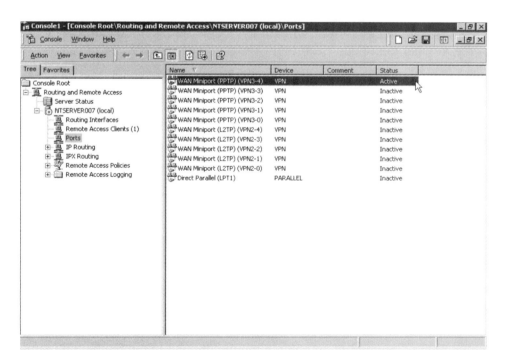

Figure 6.46   Viewing the RRAS ports object to verify the VPN connection.

Thus, the Details tab can be used to provide a summary of the parameters used for your VPN connection.

Previously this author mentioned a popular adage frequently referred to by one of his former professors. In concluding this chapter, let us return to the Routing and Remote Access screen, highlight the Ports object for our server and view the status of our VPN ports. Figure 6.46 illustrates the fact that one of the ports we previously configured and whose former status was inactive is now active.

While there are obviously a large number of parameters you need to consider to support different VPNs with applicable restrictions commensurate with organizational requirements, for most organizations you can use existing Microsoft tools, such as policy settings and group memberships to simplify the configuration process. Only when remote users are each communicating with a server via individual Internet connections, such as cable modem or DSL connections, and each user requires an individually tailored account, will the support of VPNs for a large number of users become time-consuming. Because the security afforded by VPNs by far outweighs the effort involved in configuring clients and servers, this technology should be at the top of network manager's and LAN administrator's 'to do' lists when examining actions that can be taken to secure communications occurring over the Internet or even when you wish to secure wireless transmission on a corporate intranet.

chapter seven

Service Provider-Based VPNs

Until this chapter our primary focus was upon understanding various underlying VPN technologies and how they can be used to create a virtual private network. While many organizations have resources to include personnel and funds necessary to acquire necessary hardware and software as well as to create and operate their own VPNs, other organizations may have constraints that make the use of service provider VPNs more appealing and that is the focus of this chapter. In this chapter we will turn our attention to service provider VPNs, first noting the rationale for their use. Once this is accomplished we will look at obtaining an understanding of existing and emerging transport facilities being offered, or expected to be offered, by VPN service providers. Until now our primary focus has been upon organizational methods to construct VPNs without focusing attention upon carrier transport facilities. Because many service provider VPN offerings are tied to a particular transport facility we will examine common offerings in this chapter.

Because placing the management of an organizational VPN into the hands of a third party can be risky, network managers and LAN administrators rightly want some yardstick or measurement capability associated with communications metrics that denotes a level of expected performance and penalties if acceptable performance is not forthcoming. That yardstick is more formerly referred to as a service level agreement (SLA) and will also be discussed in this chapter.

Once we obtain an idea of the rationale behind using service provider-based VPNs, carrier transport facilities, and their measurements, we will then go on a brief tour of some popular offerings. Because the operation of service provider-based VPNs can vary because of the transport technology used by the vendor, we will also note the variations associated with different service provider-based VPNs resulting from differences in their underlying technology.

Virtual Private Networking   G. H. Held
© 2004 John Wiley & Sons, Ltd   ISBN: 0-470-85432-4

While most organizations may currently consider the use of VPNs for their data transport, we will also note how differences in the underlying technology used to support VPNs can have a bearing on the suitability of the VPN for conveying voice.

7.1 Rationale for use

There are a range of issues network managers and LAN administrators face when creating a VPN. Some issues may be administrative in nature, such as deciding who will be responsible for software updates and equipment installation at various VPN locations. Other may be of a technical nature, ranging from deciding upon the type of VPN to establish, upgrading software and acquiring applicable hardware, to configuring various products. Needless to say, such activities can stress the IT budgets and personnel resources of many organizations. Recognizing this fact, several communications carriers have introduced managed VPN services in the past few years, with additional vendors announcing plans to enter this area when this book was written.

Table 7.1 lists the major reasons most organizations consider when attempting to evaluate a local VPN construction project versus the outsourcing of a VPN capability to a third party. As you might expect, the usual suspects of economics and personnel limitations are at the top of the list, however, as we discuss each reason we will note that in many instances a variety of issues based upon the structure and operation of an organization can tailor the rationale either positively or negatively.

7.1.1 Economics

Depending upon the current method of communications, cost may or may not be a driving force for establishing a service provider based VPN. For example, if your organization is using dial-up remote access via PPTP to a centrally located Microsoft Windows server, it is more than likely that any VPN service provider will charge your organization considerably more than what you are paying for essentially a dial-up remote access technology. However, if your organization is considering a leased line VPN or using frame relay or the Internet to establish high-speed connections between geographically separated locations, costs could be much more favorable for a service provider solution. In this situation, your organization might be able to avoid the need to hire a VPN or network security administrator, whose salary without benefits can range between $60,000 and $80,000 per year. In addition, site visits, other travel expenses, as well as education and training required to keep

TABLE 7.1 Rationale for a managed VPN

Reason	Description
Economics	Often the cost associated with a managed VPN can be less than the do-it-yourself approach.
Personnel limitations	The ability to recruit or have on-board trained specialists may be limited.
Reliability	A service provider may be able to replace failed equipment or re-route inoperative circuits faster than an organization can.
Communications unity	The service provider may be able to use existing T1 access line circuits or other organizational communications facilities for voice, data and VPN services.
Management	Managed VPNs may result in statistics concerning the use of the VPNs that indicate performance as well as potential bottlenecks.
Installation and support	A service provider may be able to interconnect a number of remote locations via VPNs faster than existing organizational personnel levels could.
Packaged security	Some service providers bundle a firewall, anti-virus software and optional backup software that forms a one-stop security purchase capability.

up-to-date with the technology represent costs that could be significantly reduced or eliminated from the bottom line when a VPN service provider is used. Thus, there are a range of costs an organization needs to consider when evaluating the economics associated with the use of a VPN service provider.

7.1.2 Personnel limitations

Until the Internet bubble burst, wages of IT professionals were increasing at a rapid rate. Even in a more benign economy, it is often difficult to hire personnel with a background in the technologies associated with the type of VPN you are considering for your organization. In addition, if your organization has offices in less than desirable locations, it may be difficult to hire trained personnel to work at those locations.

Because most organizations establishing VPNs are extremely security aware, it makes little sense to acquire hardware and software and skimp on your training budget. In addition, you need to recognize that security training represents a continuing education process and in-house personnel will more than likely need to attend security-related courses and seminars on a periodic basis to keep abreast with this rapidly changing field. While it is possible for many large organizations to successfully budget for the previously mentioned training, when personnel attend such training they are obviously not available to provide organizational support. When you contract with a VPN service provider, that organization is usually of sufficient size that they can provide additional support when their employees that are assigned to your organization are attending training or on a vacation or extended sick leave, ensuring your organization always has technical support. Thus, there are a range of personnel issues to include support when employees attend training, or are on vacation, that need to be examined when you consider the use of a VPN service provider. While in many situations the VPN service provider can be expected to provide a satisfactory answer to your concerns, this may not always be true and information about personnel coverage needs to be determined prior to issuing a contract.

7.1.3 Reliability

There are several areas of reliability an organization needs to consider when examining the potential use of a VPN service provider. While hardware and software reliability is obviously important, organizations need to know how a VPN service provider will react to the failure of communications. In addition, because a long distance operator may not be responsible for the local access line, it is important to differentiate between the local loop and the backbone network when discussing reliability.

If the service provider is not responsible for end-to-end transmission, it is still possible for the provider to offer a higher level of reliability than if your organization operated your own VPN. This results from the fact that many VPN service providers are communications carriers that operate large-scale nationwide or international networks. Such VPN service providers obviously know their networks and can more than likely expedite the re-routing of a backbone failure better than if your organization contacted them after a failure was observed. In addition, most communications carriers that provide VPN-managed services operate a mesh structure network. Because a mesh structured network by design has alternate routing capability, this fact alone adds to the reliability offered by the service provider in comparison to a VPN created by an organization through the use of leased lines.

For the situation where the service provider does not control the local loop, the potential for line restoration in the event of a problem may be higher than if your organization operated the VPN. This results from the fact that many VPN service providers that are also communications carriers operate very sophisticated network management centers. When they discover a local loop outage they are able to immediately contact the appropriate local service provider and arrange for service more rapidly than if your organization contacted the local service provider directly.

While VPN service providers that are communications carriers can usually provide an enhanced level of reliability, it is also possible for other types of VPN service providers to offer a similar degree of reliability. This is because many large international organizations that perform outsourcing to include EDS and IBM have extensive relationships with one or more communications carriers. Thus, you should not dismiss VPN service providers that do not directly operate communications networks as their relationships with conventional carriers could result in the level of reliability equivalent to that offered by service providers whose main business is in the communications carrier area.

7.1.4 Communications unity

If you looked at the branch offices of many organizations, you would note they operate several types of communications facilities. In addition to a voice network that may include a leased line to the branch PBX, the branch office may also have another leased line to transport data. Then, separate transmission facilities could exist to support dial-up remote access and Internet connectivity or the office may use one or more additional leased lines for those applications. With the integration of voice and data networks providing considerable economic advantages, many organizations took steps at the turn of the millennium to add voice and data multiplexes to their branch offices. In addition, the benefits of voice and data convergence resulted in other organizations implementing voice over IP, enabling voice and data to be transported over a common network infrastructure.

If your organization has previously implemented voice and data convergence or is planning to do so, the addition of a VPN application can occur alongside achieving the goal of communications unity. This is because VPN transmissions in many instances can occur over available bandwidth on an organization's current access lines, such as T1 and fractional T3 or full T3 transmission facilities. Thus, another benefit of VPNs offered by a service provider is that they enable your organization to maintain communications unity. In addition, because a service provider-based VPN might negate the necessity of remote access VPNs using dial-up transmission, the effect of the service provider can be to consolidate the use of organizational transmission facilities.

7.1.5 Management

The old adage that 'pioneers get arrows in their backs' is also true concerning the use of VPN service providers. If we assume your organization is not a pioneer that is signing the first VPN contract to be obtained by a service provider, there is an extremely high probability that the service provider is performing a similar function for many organizations. This means that the service provider more than likely has developed a range of statistical analysis tools that will be used to manage your organization's VPNs as well as that of other customers. Such information is then used to promote the management of your managed VPN facilities to include observing utilization, delay and other metrics that are valuable for accessing the level of service provided to your organization.

7.1.6 Installation and support

Because most VPN service providers are larger than the companies they support, they can normally be expected to have a staff trained in the installation of VPN hardware and software. Because their personnel more than likely have performed several installations, they should be literally higher on the training curve than if your organization used in-house personnel to set up VPN hardware and software. This ability to expedite the installation of hardware and software can be an important criterion when your organization is implementing a new project that depends upon the availability of a VPN capability at a fixed point in time.

Once a VPN is operational, the level of support becomes an important metric to consider. For example, how responsive will the service provider be to a failure occurring after normal business hours or on the weekend? Will the level of response be similar in Boise, ID, and Philadelphia, PA, or can you expect differences in the level of response based upon different organizational locations? If the VPN service provider is a national or international organization, you have a realistic expectation that their level of support will be beyond that of most organizations that take the do-it-yourself route and construct their own VPN. However, as President Reagan once said, 'Trust is one thing, verification is another.' While it is highly doubtful if President Reagan knew anything about VPN technology, his words provide an important lesson when considering the use of a VPN service provider. That is, make sure that your contract documents installation dates and response times for support and includes either penalties to the provider or an exit from a contract if the level of support does not live up to your written expectations. As we discuss service level agreements later in this chapter, we will note why they may require a bit of negotiation and modification to protect organizations from the unexpected.

7.1.7 Packaged security

Another area worth noting when considering the use of a VPN service provider is the level of security they provide. While some vendors limit security to VPN hardware and software, other vendors provide a packaged approach to security. Concerning the latter, such vendors may include installing virus-scanning software on a mail server and individual PCs, management of router access lists and firewalls, and the operation of intrusion detection software, as part of a security package bundled with the VPN. Because many VPNs are IP based, a security package makes sense since the IP VPN may occur over a common access line, which also provides the employees of the organization with access to the Internet. By providing a security package to include taking responsibility for updating equipment configuration settings and virus detection software as well as monitoring the operation of intrusion detection software, the VPN service provider in essence begins to resemble a home or office security alarm monitoring service. By charging a nominal fee for each customer location, the VPN service provider can often provide security monitoring and necessary responses to attacks for a fraction of the cost of organizations that build their own VPN and use in-house personnel to provide necessary security related functions.

Now that we have an idea of the main reasons many organizations use a VPN service provider, let us turn our attention to the transport facilities they may offer. In doing so we will note that there are differences in the transport facilities that can result in variances in support of VPNs transporting time-sensitive information, such as voice and video.

7.2 Transport facilities and VPN operation

Until a few years ago it was a relatively simple process to note the type of transport facility offered by a VPN service provider and equate that facility to a particular type of VPN. For example, layer 2 frame relay-based VPNs were usually configured via the use of frame relay switches that formed the backbone of a service provider's network. Over the past few years the addition of ATM switches to the backbone infrastructure of many communications carriers has resulted in both frame relay and IP being transported over ATM. Depending upon the ends of the VPN tunnel, the transport of data over the ATM infrastructure may represent only a portion of the backbone network used for end-to-end communications. To further complicate matters, there are several existing and an emerging VPN technology that provide a variety of choices to the organization seeking a managed VPN network. While frame relay, IP variations such as PPTP, L2TP and IPSec VPNs are offered by many service providers, other providers are beginning to offer VPNs based upon Multi-Protocol Label Switching (MPLS). Thus, similar to the statement of Barry Goldwater when he ran for US President in 1964, you have 'a choice, not an echo' when you are considering different VPNs offered by service providers. In this section we will examine several hardware- and software-based VPN technologies, noting how they operate and the advantages and disadvantages associated with each technique. To simplify our discussion we will subdivide the communications carrier backbone infrastructure into hardware- and software-based switching.

7.2.1 Hardware-based switching

Hardware-based switching involves the use of frame relay or ATM switches to provide the backbone transport facility for VPN tunnels created over the backbone. Figure 7.1 illustrates the flow of data over a hardware-based switching infrastructure consisting of ATM switches. In this example an IPSec VPN is shown overlaid on the ATM backbone.

If your organization was creating a frame relay VPN over the ATM infrastructure, the router could be replaced by a Frame Relay Access Device (FRAD) and due to the lack of an IP transport the need for a firewall would more than likely disappear. The firewall would disappear if your organization only used frame relay on the access line and avoided any public network connection.

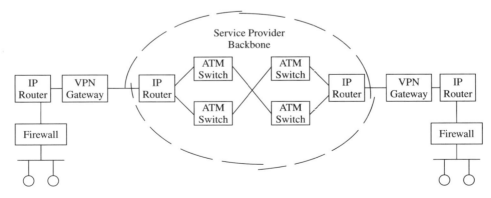

Figure 7.1   Creating an IPSec VPN over a hardware-based switching network.

Because a frame relay network based upon pre-defined permanent virtual circuits (PVC) represents a fixed, closed network, your organization would be isolated from other traffic and would not require the use of a firewall. However, this would not be true if frame relay access occurred via IP and your organization also used a common access line for Internet access.

The key advantage associated with the use of hardware based switching is a reduction in packet transmit time. This results from the fact that switching occurs based upon an examination of cell or frame headers by the hardware, which is significantly faster than software based switching. Although the use of a hardware-based switching infrastructure is better able to support minimal latency applications, such as voice and video, it is important to understand exactly how the communications carrier will handle your organization's VPN traffic to determine if the carrier can configure its backbone to provide the latency required for voice and video. As an alternative, some VPN service providers will denote end-to-end latency as part of their SLA.

7.2.2 Software-based switching

There are two types of software-based switching that can be used by communications carriers to provide a VPN transport capability. Those software-based switching methods are MPLS and IP, with each method based upon MPLS or IP switching software being used in the carrier's backbone routers.

In reality, there are two types of MPLS VPN technology: layer 2 and layer 3. Layer 2 MPLS VPNs can be considered as an emerging technology, which is based upon an Internet Engineering Task Force (IETF) draft RFC. Under layer 2 MPLS two labels are prefixed to network traffic. One label represents a point-to-point virtual circuit, while the second label represents the tunnel across the network. Although a majority of layer 2 MPLS VPNs are expected to occur by the encapsulation of Ethernet, the technology can also be used to support frame relay, ATM, HDLC and the Point-to-Point Protocol (PPP).

Once the layer 2 traffic is encapsulated, the local Label Switch Router (LSR) connected to the network assigns a virtual circuit label, which identifies the VPN or connection end point. This label is equivalent to a Frame Relay Data Link Connection Identifiers (DLCI) used to create a permanent virtual circuit for routing of data over a frame relay infrastructure. At the entrance to the providers network another LSR uses the virtual circuit label to determine how to process the frame and then adds the tunnel label that defines the route the data takes through the network.

One key advantage associated with a layer 2 MPLS-based VPN is the ability to directly map 802.10 virtual LANs across a wide area network, maintaining

802.1p priorities. This permits service providers to offer QOS guarantees needed to support both voice and video via a VPN. Another key advantage associated with layer 2 MPLS VPNs is the fact that they are circuit based. This means that the flow of data over the provider's backbone is as secure as that of a connection-oriented technology, such as frame relay or ATM. However, this security only holds true for traffic routed through the carrier's backbone. If your organization uses a common access line for VPN traffic as well as regular Internet access, you will need to consider the use of router access lists and a firewall to protect each terminal point of the VPN.

In contrast to a layer 2 MPLS VPN, layer 3 relies on IP routing to construct paths. In doing so only one label is appended to data, which is used to route traffic through the network. Figure 7.2 illustrates the flow of IP traffic over a router-based MPLS network.

IETF standards define the manner by which MPLS layer 3 VPNs support Differentiated Services, permitting network service providers with the ability to map ingress priorities assigned in the IP header to a priority through the MPLS network.

In comparing the use of a switch-based infrastructure to a router-based infrastructure, the switch-based infrastructure will normally provide a lower level of latency through the network. However, because IP flowing over ATM requires packets to be mapped to cells, some of the savings associated with the use of ATM switches are negated by the need to deconstruct and then reconstruct packets. In addition, because routers inherently have a more sophisticated routing capability than switches, it is possible to create an interconnected mesh structured network using routers that provide a higher degree of alternate facilities in the event of equipment failure. However, because MPLS occurs via software, even though the goal of the technology is to expedite traffic through the network, it will not flow as rapidly as via an ATM backbone. Another difference between hardware- and software-based switching lies in the fact that ATM provides the ability to obtain a guaranteed

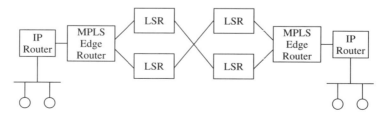

Figure 7.2   The flow of IP traffic via MPLS occurs on a router-based network.

QOS capability. Thus, communications carriers that provide a VPN service over ATM switches can be expected to be able to provide a better capability to support voice and video applications than their competitors that offer a VPN creation capability over a software-based switching infrastructure. Concerning VPN capabilities, we can obtain an appreciation for how they are defined by turning our attention to Service Level Agreements (SLAs).

7.3 Service level agreements

While the term service level agreement or SLA is familiar to most readers, it is probably a good idea to define this term. A Service Level Agreement can be considered to 'represent a binding contract between a customer and a network service provider.' This contract usually specifies in measurable terms the services that the network provider will furnish as well as the recourse to the customer if one or more metrics do not meet or exceed the values specified in the contract.

A word of caution is in order concerning customer recourse if the level of service of the service provider is not what you expect. Most SLAs have a monetary credit provided to the customer when the level of service falls below a certain metric. While receiving a credit for poor service may appear to right a wrong, the actual credit may be minuscule. This is because credit for poor performance is usually based upon a 24-hour clock. That is, if performance is severely degraded during a one-hour period your organization might receive credit for only 1/24th of the service charge for one day, even if the one-hour period was the busy hour during the heaviest traffic day of the month. A second problem associated with some SLAs is a cap or limit on the amount of credit that can be applied to an organization's monthly bill or in some cases, over the life of the contract. Unless there is a customer exit clause in the contract that allows your organization to take its business elsewhere, you may either have to endure an unexpected level of poor performance or accept a penalty for being able to exit the poor level of services provided by a service provider. Even though the probability of receiving a continuously poor level of performance is minimal, it is always a good idea to have an exit clause within an SLA.

Now that we have an overview of the basic SLA, let us discuss the metrics that can be incorporated into an SLA, with particular attention focused on VPN SLA metrics.

7.3.1 SLA metrics

There are a wide range of metrics that can be included within an SLA, with the actual specified items dependent upon the type or types of VPNs the service

TABLE 7.2 Common Service Level
Agreement elements

Availability
Connectivity Guarantee
Dial-in access availability
Number of simultaneous users
End-to-end latency
Packet delivery
Hardware resolution
Notification of network changes
Help-desk response time
Usage statistics

provider offers to your organization. Table 7.2 lists ten typical SLA elements that are quantified by one or more metrics within the written agreement. In the remainder of this section we will discuss each element and the metric or metrics that are commonly used to quantify each element.

7.3.1.1 Availability

An availability guarantee normally refers to the percentage of time the provider's network is operational and available for use. Most service providers that operate modern digital networks typically offer an availability guarantee in the high 99 percent range.

7.3.1.2 Connectivity Guarantee

Care should be taken not to confuse availability with a connectivity guarantee. Whereas availability indicates the percentage of time the provider's network is available for access, by itself this does not mean an organization employee who gains access to the network will be routed to the destination they desire and make a connection to the distant end. End-to-end connectivity is normally offered by a connectivity guarantee, which is expressed as a percentage. Because modern carrier networks are based upon the use of hardware- or software-based switches, the flow of traffic through a network switch towards another device is limited. Due to this, some service providers may specify a maximum number of simultaneous users being able to obtain end-to-end connectivity instead of a connectivity guarantee.

7.3.1.3 Dial-in Access Availability

Although your organization may be using a service provider to connect two or more LANs at geographically separated locations together via a VPN, chances are rather high that employees on travel status will need to connect to your organization's computer resources. When planning for this situation you need to consider dial-in access availability as well as the type of VPN connection to your organization's fixed locations the service provider will support. Dial-in access availability can occur via the use of private telephone numbers to include 800-type access or via an Internet connection.

7.3.1.4 Number of Simultaneous Users

As previously discussed, an alternative to a connectivity guarantee occurs when a service provider guarantees a maximum number of simultaneous users having the ability to reach their VPN destination. When working with an SLA element based upon the number of simultaneous users to be supported, it is important to compare what is offered by the provider to your organization's operation. For example, if the service provider will guarantee service for 50 simultaneous users and your organization has 100 users, it is important to note the activity of those users. If a maximum of 35 users require simultaneous access to the VPN service, then the service guarantee should be more than acceptable. However, if your organization will have more than 50 simultaneous users, the service provider only guarantees to support 50. While it is possible your users may never experience a problem, it is also possible that your users could constantly encounter problems. If so, as long as such problems result from more than 50 simultaneous users seeking VPN support, your service provider is living up to the terms of the SLA to the chagrin of your user population.

7.3.1.5 End-to-end Latency

There are several methods commonly used by service providers to specify end-to-end latency. First, latency is specified either in terms of customer provided equipment (CPE) to CPE through the carrier's network or from an ingress point to an egress point in the carrier's network. Obviously, the latter does not include delays attributable to CPE and each access line. Thus, for most organizations, a far better measurement concerning latency is from CPE to CPE.

Most latency measurements are defined within an SLA as the average packet round trip delay. This delay time will be defined based upon the location of customer sites. For example, a service provider might specify one VPN latency commitment for customer sites within North America, while other latency commitments could be specified for sites within Europe and sites within North America that access Europe and vice versa. Depending upon your organization's requirements for the transport of real-time data, such as voice over IP and video over VPN connections, the differences between vendor metrics defined for this element may or may not be meaningful.

7.3.1.6 Packet Delivery

While it is important for packets to reach their destination with a minimum of delay, it is more important that packets reach their destination. Packet delivery denotes as a percentage the ability of transmitted packets to reach their destination. Because routers are designed to discard packets under extreme loading conditions, the willingness of a service provider to offer an extremely high level of packet delivery can be viewed as an indicator that the service provider's network has sufficient capacity to support customer requirements over the life of the contract.

7.3.1.7 Hardware Resolution

When a service provider either takes over responsibility for existing equipment located at your organizational offices or installs new equipment, you need assurance that any hardware problem will be expediently resolved. Some service providers will only define a hardware resolution period for equipment they provide, while other service providers may include existing equipment within their guarantee. For both situations a hardware resolution period is usually specified in terms of hours, with the clock commencing only after the service provider is contacted and the problem verified by their technician.

7.3.1.8 Notification of Network Changes

One element normally not included within many SLAs but nevertheless important is advanced notification of network changes. Such changes can include a reconfiguration of communications carrier equipment within the backbone of their network or the upgrade of software on Customer Provided Equipment (CPE) located at your organization. For either situation most organizations appreciate advanced notification as well as the willingness of the service provider to coordinate possible changes to their schedule of network changes that minimizes the effect upon customer activities.

7.3.1.9 Help-desk Response Time

While all VPN service providers offer a help-desk facility, not all facilities have the same level of personnel expertise. Thus, a help-desk response time element within an SLA may not be meaningful. For example, a communications carrier that specifies that their help-desk responds to customer queries within a 15-minute average does not indicate if the response was correct and if the problem was solved. Unfortunately, at the present time, communications carriers that include a help-desk response time element in their SLA only define a metric based upon the duration of time required for getting back to the customer and do not include metrics that indicate the actual level of help-desk response.

7.3.1.10 Usage Statistics

Most VPN service providers will furnish a wide range of usage statistics to their customers. When included within an SLA, the provider usually indicates the date within a month whereby statistics for the previous month are provided to the customer. Thus, when examining an SLA that includes usage statistics, it is important to determine what type of statistics are provided and when they are provided. In addition, customers should also investigate the manner by which statistics are provided. In the era of the Internet, many service providers offer customers essentially real-time access via the Internet to usage statistics. Customers can retrieve usage statistics for pre-defined periods of time ranging from hours to days and weeks or months. In addition, the statistics offered by some vendors let the customer's click of a mouse compare usage for different periods of time. Because usage statistics provide network managers and LAN administrators with an indication of the usage of the VPN maintained by the service provider, they can allow the customer to review the managed network. In addition, by tracking different metrics, such as packets sent and latency over a period of time, you may be able to spot potential bottlenecks prior to such bottlenecks materializing.

7.3.2 SLA limitations

While SLAs are important to consider, they are not problem-free. Perhaps the biggest problem with respect to SLAs is the utility of metrics when a VPN crosses autonomous systems operated by another carrier, such as an IP VPN, which flows across the backbone of another carrier. While a service guarantee in a single domain is relatively easy to define, when a VPN crosses a public peering point, there is usually no control for the VPN pipe at that type of peering point. In comparison, if the VPN pipe crosses into another carrier's network at a private peering point it becomes possible for the two carriers to negotiate the manner by which the VPN pipe is supported through the third party network. Although the routing of a VPN through one or more third party networks may not be directly known to your organization, it is usually reflected in the scope of SLA metrics offered by a service provider. That is, when the service provider is not in full control of end-to-end transmission, they may need to add a literal 'fudge factor' to their metrics to account for situations beyond their control. Of course, when the carrier uses the transport facilities of the public Internet for a portion of network data delivery, the cost should be less than when their internal network is used for end-to-end network delivery.

Now that we have an idea of SLAs that includes problems when data flows over multiple networks, we will conclude this chapter by turning our attention to VPN services offered by several communications carriers.

7.4 VPN service provider overview

One of the problems associated with a discussion of communications carrier VPN offerings is the high level of turmoil in the industry. During the book writing process this author observed the bankruptcy of Genuity, PSInet, WorldCom, now renamed MCI, and Global Crossing as well as the admission that WorldCom had literally cooked its books to the tune of in excess of $10 billion and was being referred to by some persons as FraudCom. Thus, in this section I will provide a brief overview of several communications carrier VPN services and recommend that readers contact those and other carriers to ascertain current offerings. In addition, due to the over-capacity and literal glut of optical fiber, many communications carriers are heavily discounting VPN multi-year contracts. Due to this, cost is extremely variable and will not be discussed in this section.

As you might expect, a majority of the largest VPN service providers are also communications carriers. According to published reports, AT&T, Sprint and the now bankrupt WorldCom that was renamed MCI accounted for over 40 percent of the VPN market, with the remainder distributed over numerous national and international organizations to include Infonet, Cables & Wireless, Level 3 Communications which acquired Genuity in February 2003 and other companies to include Equant which resulted from the merging of GlobalOne and Equant during 2000 and which has more than half its shares owned by France Telecom. To obtain an overview of VPN service provider offerings at the time this book was prepared we will turn our attention to a brief examination of the offerings of AT&T, Level 3 Communications, Sprint and Verizon.

Concerning Verizon, although this company was not offering VPN services when this book was prepared, it had announced its intention to do so. Since Verizon has sales of approximately $50 billion per year, it was felt that a brief review of that carrier's intentions is warranted as it operates an extensive IP network and could become a major VPN service provider within a few years.

7.4.1 AT&T Corporation

AT&T Corporation provides a global suite of network and premises-based IP VPNs. Customers can select from a range of transport, network access and security options that in effect provide the ability to customize a solution to an organization's requirements.

Currently AT&T provides frame relay, ATM and IP transport options. Network access options include remote dial-up, dedicated, broadband and DSL. Concerning security, AT&T can support managed firewalls, intrusion detection and network scanning.

One area which separates AT&T-managed VPN services from some competitors is the scope of its international network, which interconnects over 180 countries. In addition, the well-known AT&T Operations desk can support customer-managed VPN services to include receiving alarms, generating trouble tickets and providing problem resolution notifications. Another area of AT&T VPN services that warrants attention is its MPLS-based network environment, which permits VPN traffic to be routed similar to a circuit switched environment, which enhances security while permitting a defined QOS.

One of the more interesting aspects of AT&T's IP network is the public's ability to directly view its performance. Figure 7.3 illustrates the AT&T IP Network Performance Web page during February 2003. Note that at the time the Web page was accessed the AT&T IP network had an average round-trip delay of 35 ms and zero packet loss. By clicking on the entry in the far left column of the Web page you can obtain specific information about current network performance which is shown displayed in Figure 7.3, backbone delay and loss information, the prior month's network performance averages and network status information.

One most interesting AT&T performance screen is the network performance page. You can select the display of performance metrics from any of the 18 US cities to the other 17 cities, resulting in the display of a color-coded map in the top half of the page which is followed by a table that summarizes the results for the selected city in the lower half of the page. The lower half of the AT&T current network performance page is shown in Figure 7.4. Note that the results are shown from Atlanta to 17 other cities in the United States, with the table indicating the round-trip delay in ms from Atlanta to each city as well as packet loss.

Figure 7.3 The general public can access a variety of AT&T IP network performance metrics via the Internet.

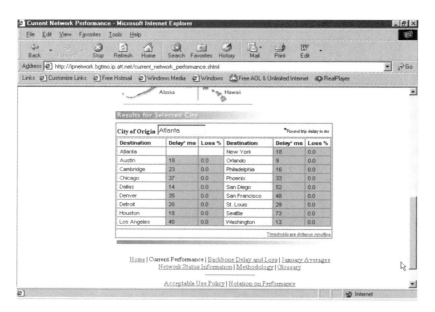

Figure 7.4 The AT&T Network Performance display permits the general public to view round-trip delays and packet loss from a selected city to 17 additional locations.

At the time the table data was captured, AT&T was not experiencing any packet loss on their network, while round-trip delays could be generally equated to the distances between locations.

AT&T offers customers several VPN solutions, to include IPSec, L2TP and PPTP. For dial-up VPN access, AT&T now offers an SLA that includes a 99 percent call success rate and provides a variety of Service Level Reporting methods to its VPN customers, enabling them to monitor the service performance of their virtual network.

7.4.2 Level 3 Communications

During February 2003 Level 3 Communications acquired the assets of Genuity, which operated a comprehensive Internet backbone but was bankrupt. Level 3 Communications markets VPN Advantage (VPNA), a managed VPN service based upon the use of Nortel Networks hardware.

Marketed as a lower cost alternative to a private line network, VPNA represents a service that transports data over the former Genuity IP network backbone in encrypted form. VPNA provides customers with $24 \times 7 \times 365$ management, monitoring, maintenance and response services that are backed by defined service level agreements.

The Genuity network acquired by Level 3 Communications spans approximately 20,000 miles, connecting 33 metropolitan networks in North American and Europe. The predecessor to Genuity, BBN Corporation, established the world's first packet-based VPN in 1975, resulting in Level 3 Communications' acquisition of Genuity including almost 30 years of VPN expertise. In addition to offering a managed VPN service, Level 3 Communications supports L2TP in dial networks which enables remote access to managed services.

7.4.3 Sprint

Although Sprint is usually viewed as the smallest of the big three communications carriers, it has several interesting VPN capabilities that deserve mention. First, Sprint can arrange for users of other service providers to securely interconnect to its customer VPNs, in effect, providing an extranet VPN capability. This can be an important consideration when an organization needs to provide an economic and secure method of access from its suppliers or customers. A second interesting aspect concerning Sprint's VPNs is its view toward equipment. Sprint permits customers to purchase or rent VPN-compliant products. Sprint product offerings include routers, firewalls, CSU/DSUs and authentication servers, such as RADIUS and security token-based servers.

Until mid-2003 Sprint VPN offerings fell into three general categories – SprintLink Frame Relay, SprintLink Packet Private Line and SprintLink Virtual LAN Services. Under SprintLink Frame Relay customers operate a frame relay User Network Interface to the edge of Sprint's network, where frame relay Data Link Connection Identifiers (DLCIs) are mapped to L2TP tunnels running over Sprint's backbone. While the customer can continue to use frame relay equipment, in effect the VPN occurs over IP in the Sprint network.

A second Sprint VPN offering, SprintLink Packet Private Line, emulates the operation of a circuit, providing customers with what appears to be a full or fractional circuit speed over Sprint's IP network. The third Sprint offering, SprintLink Virtual LAN Services, represents an unmanaged set of services that enables customers to create a global Virtual LAN via the 802.1Q Ethernet VLAN protocol to interconnect geographically separated locations.

In mid-2003 Sprint announced that they would offer an MPLS-based VPN service by year-end. Until then, Sprint considered MPLS as a solution seeking a problem and considered the availability of bandwidth on the carrier's backbone as eliminating the need for MPLS. Although to this author's best knowledge no customers complained to Sprint about any latency problems, the carrier's willingness to offer MPLS appears to result from competitive offerings rather than necessity.

For each VPN method, Sprint SLAs cover network availability, backbone delay and include a busy-free guarantee. The actual SLA metric is based upon the type of access to the Sprint backbone, with a 100 percent availability guarantee offered to customers that are in Sprint's Broadband Metropolitan-area SONET Network (BMAN), while the availability guarantee is reduced to 99.9 percent when a traditional leased line access to the Sprint network is used.

Similar to other service providers, Sprint offers $24 \times 7 \times 365$ technical support. Sprint also provides access to repair technicians for VPN devices it manages.

7.4.4 Verizon

Formed from the merger of GTE and Bell Atlantic, Verizon was in the process of gearing up to offer managed VPN services when this book was written. Although Verizon previously offered VPN services through its once closely connected IP partner Genuity, it was in the process of rolling out its own services since Genuity was acquired by Level 3 Communications.

Verizon was expected to use Nortel's Contivity and Cisco Systems' 3000 series devices to form the infrastructure for providing managed VPN services.

Because Verizon's IP network is not a true national backbone, the company negotiated an agreement with Sprint to use its IP network to support customers in areas where its own IP network lacks coverage. Verizon is also in the process of constructing an MPLS network over its own IP network, permitting the carrier to support layer 3 VPN services by dedicating paths over its network for individual customers.

VPN Checklist

The selection of an appropriate VPN solution requires taking a large number of parameters into consideration. Those parameters can include the location of employees who need to communicate in a secure manner over a public network as well as numerous technical features that need to be carefully matched against organizational requirements.

Because organizational requirements can and usually do differ from one organization to another, there is no one best VPN solution. However, by considering the range of parameters and technology associated with VPNs it is possible to match a particular VPN solution to organizational requirements. In this appendix readers will find a checklist of parameters and features they may wish to match to their organizational requirements. Because it is also possible to turn to a third party to provide a VPN solution, by carefully examining and selecting those features required by your organization from the following checklist, this list can also form the basis for the creation of a Request for Proposal (RFP) to solicit bids from vendors.

The VPN checklist presented in this appendix can also be used to compare and contrast vendor offerings by matching the requirements you select against different products. In addition, because it is possible that an alternative to a VPN may be in the best interest of your organization, this checklist concludes with a list of alternatives you may wish to explore.

VPN Feature/Parameter	Organizational Requirement
VPN category	
Site-to-site	_____
(under IPSec called gateway-to-gateway)	
Remote access	_____

Virtual Private Networking G. H. Held
© 2004 John Wiley & Sons, Ltd ISBN: 0-470-85432-4

VPN Feature/Parameter	Organizational Requirement
(under IPSec called host-to-host)	
Combined site-to-site/remote access	_____
Authentication	
Password Authentication Protocol (PAP)	_____
Challenge Handshake Authentication Protocol (CHAP)	_____
Microsoft CHAP (MS-CHAP)	_____
MS-CHAPv2	_____
Extensible Authentication Protocol (EAP)	_____
Use of EAP	
Token card	_____
One-time password	_____
Public key	_____
Digital certificates	_____
Other	_____
EAP-Transport Level Security	_____
(for mutual authentication support)	
Encryption and hashing	
Data Encryption Standard (DES)	_____
Triple DES	_____
Advanced Encryption Standard (AES)	
128 – bit key	_____
192 – bit key	_____
256 – bit key	_____
Public key utilization	_____
Hash algorithms	
Secure Hash Algorithm 1 (SHA1)	_____
Message Digest 5 (MD5)	_____
Software operating system support	
Windows version(s)	_____
Unix version(s)	_____
Linux version(s)	_____
Other	_____
VPN protocol	
Frame relay	_____
Point to Point Tunneling Protocol	_____
Layer 2 Forwarding (obsolete)	_____
Layer 2 Tunneling Protocol (L2TP)	_____

VPN Feature/Parameter	Organizational Requirement
L2TP with IPSec	_____
IPSec	
Authentication Header (AH)	_____
Encapsulated Security Protocol (ESP)	_____
Internet Key Exchange (IKE)	_____
Mode of operation	
Transport mode	_____
Tunnel mode	_____
Secure Sockets Layer (SSL)	_____
SSL with network appliance	_____
Vendor provided	
Type	_____
Redundancy	_____
Scalability	_____
Service Level Agreement	
Availability	_____
Connectivity guarantee	_____
Dial-in access availability	_____
Number of simultaneous users	_____
End-to-end latency	_____
Packet delivery	_____
Hardware support	_____
Notification of network changes	_____
Help desk response time	_____
Usage statistics	_____
Interoperability	
With specific vendor (s) product (s)	_____
With installed product(s)	_____
With anticipated product(s)	_____
Alternatives to explore	
Modem dial-up	_____
ISDN utilization	_____
Switched digital service	_____
Dedicated leased lines	_____
Project collaboration software	_____
Messenger software	_____

index

l